The living tundra

Studies in Polar Research

This series of publications reflects the growth of research activity in and about the polar regions, and provides a means of disseminating the results. Coverage is international and interdisciplinary: the books will be relatively short (about 200 pages), but fully illustrated. Most will be surveys of the present state of knowledge in a given subject rather than research reports, conference proceedings or collected papers. The scope of the series is wide and will include studies in all the biological, physical and social sciences.

The living tundra

YU. I. CHERNOV

TRANSLATED BY D. LÖVE

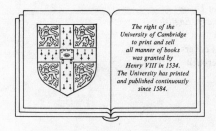

*The right of the
University of Cambridge
to print and sell
all manner of books
was granted by
Henry VIII in 1534.
The University has printed
and published continuously
since 1584.*

CAMBRIDGE UNIVERSITY PRESS

CAMBRIDGE

NEW YORK   NEW ROCHELLE

MELBOURNE   SYDNEY

Published by the Press Syndicate of the University of Cambridge
The Pitt Building, Trumpington Street, Cambridge CB2 1RP
32 East 57th Street, New York, NY 10022, USA
10 Stamford Road, Oakleigh, Melbourne 3166, Australia

English translation © Cambridge University Press 1985

First published in Russian as *Zhizn' tundry* by Izdatel'stvo "Mysl'" 1980

English translation first published 1985
First paperback edition 1988

Printed in Great Britain by
the University Press, Cambridge

Library of Congress catalogue card number: 83-26264

British Library Cataloguing in Publication Data
Chernov, Yu. I.
The living tundra.
1. Tundra ecology – Soviet Union
I. Title   II. Zhizn' tundry. English
574.5'2644'0947   QH161
ISBN 0 521 25393 4 hard covers
ISBN 0 521 35754 3 paperback

UP

Contents

Translator's foreword

The translation has been made as faithful to the original as possible. Nothing has been omitted. However a more liberal use has been made of Latin scientific names in addition to the vernacular names, especially to avoid the confusion that often arises from using only common names of plants and animals. Apart from the introduction of these Latin names, for the choice of which I myself am responsible, nothing else in the text has been altered.

The geographical names are those used in the comprehensive edition of *The Times Atlas*, and when not found there have been transliterated according to the system used therein. Some of the terms used may be unfamiliar to readers not previously introduced to the ecology of arctic areas and merit a few words of explanation, not given in the text.

'Spotted tundra' (Russian: pyatnistaya tundra) is a descriptive term for a tundra with spots of various sizes, from a few centimetres to metres in diameter, entirely or almost entirely free of vegetation. It is sometimes also called 'medallion' tundra, but I think 'spotted' is more descriptive and a direct translation of the Russian term. It is, of course, a frost phenomenon, but not necessarily associated with other types of patterned ground such as polygons, stone circles or stripes. The bare spots consist both of gravelly and clayey soils and are very wet and soggy when thawing.

'Mire' is the type of bog most common in the northlands of Eurasia where it depends more on ground water conditions than on precipitation for its existence in the, at times, relatively dry arctic and subarctic areas. The word is based on the old nordic concept of 'myr'. I have preferred to use this word in order to distinguish these 'bogs' (Russian has only one word for bog, 'bolota', whereas Scandinavia has a whole array of terms for them) from, e.g., raised bogs, which are not an arctic phenomenon. Where appropriate I have, of course, used this term as well.

'Watershed' is used herein for the more or less wide upland between two (or more) drainage areas. It is, thus, a relatively dry environment by comparison with the more waterlogged areas in the drainage basins themselves.

'Plakor' is another new term, introduced recently in the translation of V. D. Aleksandrova's *The Arctic and Antarctic: their division into geobotanical areas* (Cambridge University Press, 1980). It is derived from a Greek word for 'surface, flatland', and is used by Soviet botanists to describe a locality typical of its zone, with fine soil on level ground, well drained, neither too wet nor too dry and with snow cover melting neither too early nor too late.

The forest-tundra ecotone is defined in Löve, D. (1970). Subarctic and Subalpine: where and what? *Arctic and Alpine Research*, **2** (1): 63–73.

Most other terms are, I think, well explained in the text.

When selecting equivalent English terms I have resorted to the multi-lingual *Geobotanicheskiy Slovar'* (*Geobotanical Dictionary*) by O. S. Grebenshchikov (Nauka, Moscow, 1965) and to *Entsiklopedicheskiy Slovar' Geograficheskikh Terminov* (*Encyclopedical Dictionary of Geographical Terms*), ed. S. V. Kalesnik (Sovietskaya Entsiklopediya, Moscow, 1968).

I want to express my gratitude to Dr S. W. Greene, Department of Botany, University of Reading, England, who originally suggested this translation and who has been very helpful throughout the work. My thanks go also to Dr F. Pitelka, Department of Zoology, University of California, Berkeley, USA, for help with the names of birds, to Dr R. S. Hoffmann, Curator, Museum of Natural History, University of Kansas, Lawrence, USA, for assistance with the names of mammals, to Dr M. G. Morris, Institute of Terrestrial Ecology, Furzebrook Research Station, Wareham, Dorset, England, for help with the names of insects, to Mr M. G. Hardy, Department of Zoology, University of Reading, England, for help with the names of fishes, and Dr B. Stonehouse, Scott Polar Research Institute, Cambridge, England, who advised on the most appropriate names in a number of cases. I am also grateful to my husband, Dr Áskell Löve, for his help and encouragement, and especially to the staff at the Cambridge University Press for suggestions and support, making the work on this book a pleasure.

Doris Löve   San José, California, January 1983

Preface to English edition

Yury Ivanovich Chernov was born in 1934 into a family of village teachers and from his earliest years developed a keen sense of observation and a love for Nature, especially insects. In 1958 he graduated from the Moscow Pedagogical Institute and after post-graduate work he became at first Assistant and later Associate Professor there. He passed the examination for the 'candidate degree' (a Russian equivalent for Ph.D.) and became Doctor of Biological Sciences in 1976. In 1972 he moved to the Academy of Sciences Severtzov Institute of Evolution, Ecology and Morphology of Animals, where he is still active, directing the Laboratory of Community Structure and Dynamics. He is a highly esteemed scholar and teacher, greatly admired by all those who work with him. It is quite evident from this book that his early interest in insects, especially the Diptera, has widened to include all kinds of zoological and botanical problems, especially the interplay of living organisms. His career has made him familiar with a variety of vegetation types, ranging from semi-deserts, steppes and forests to the arctic tundra, which latter has become his favourite area. In 1976 he participated in the *Kalisto* expedition to islands in the Pacific Ocean. He is the author of several books and numerous articles and is a co-editor of the *Russian Journal of Zoology*.

The present work is his first to be translated into English. It provides a wide-ranging view of the complexities of life in the harsh environment of the Arctic tundra and pleads for conservation measures to protect the tundra ecosystem. Since it deals almost exclusively with the Russian Arctic and quotes extensively from the work of Russian scientists, it will be of interest to all who are involved with the biology of the Arctic, particularly those who are unfamiliar with Russian work because of the language barrier. It has been written for a broad spectrum of readers and should be of interest both to professional biologists and students as well as laymen, and will help to stimulate interest in a fascinating and challenging part of the world.

For the English edition an extra figure (the Map, pp. xii–xiii) has been inserted, based on the author's Fig. 1, which shows the position of localities mentioned in the text. In the index transliterated places are listed as used in the text. These are as given in *The Times Atlas of the World* (comprehensive edition, 1975), other forms are given within parentheses where it is thought necessary. One further difference between the English and Russian editions is that the originals of Figs. 17, 25, 37, 39 in the English edition are different from those used in the Russian text, as the originals of the latter were not available. The new figures were kindly provided by Dr Chernov and selected so that their subject matter is as similar as possible to those published in the Russian edition.

Finally it is a pleasure to thank Dr Löve for the care she has taken with the translation and Dr T. E. Armstrong of the Scott Polar Research Institute, Cambridge, for help with the Russian place names and the preparation of the frontispiece.

S. W. Greene, April, 1983

Abstract

The author, a biogeographer and ecologist, reviews the results of his work over many years in various parts of the Eurasiatic subarctic and analyses the classical and modern literature concerning the problems of the Soviet Northlands. On the basis of original material he scrutinizes many problems of adaptation to subarctic living conditions and analyses the structure of the fauna and flora as well as the interrelationships between groups of organisms in their special natural environment. He also discusses problems of biological production in the tundra and the approach to a rational utilization of its resources.

Map. Based on Fig. 1, showing the location of the major place names mentioned in the text.

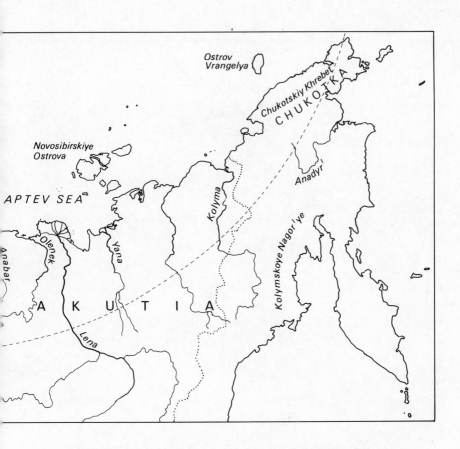

1

Introduction

With this book, first published in 1980, the author wishes to commemorate the one hundredth anniversary of the publication by Alfred Wallace of *Tropical Nature*, the brilliant and original description of life in a tropical forest.

The magnificent nature of both the Tropics and the Arctic represents, indeed, contrasting kingdoms of heat and cold. Could there be anything more different? Still, a comparison between opposites can give a clue to the understanding of complex phenomena. In the first paragraphs of his book, Wallace compared the constancy of the tropical climate with that of the ever-changing polar climates. The contrast helps us to a better understanding of the problem. Therefore a study of the tundra, the polar desert and the frozen areas is of importance not only for knowledge of the nature of the Arctic itself but also for a better general understanding of life in the biosphere.

The interdependence of natural forces is especially clear in the landscapes around the Poles. At first glance, this land is startling with its strange assortment of tundra relief – polygons, networks of stone, hummocks and plant cushions – the result of the interaction between soils, vegetation and water. The remarkable interrelationship between the tundra organisms, for instance the dependence of all species of winged and four-footed animals on the lemming, is indeed extraordinary. Under the severe arctic conditions the adaptive uniformity of the organisms to their environment is especially prominent. All this concerns many of the general principles of natural science. That is why studies of arctic Nature always reveal a wide range of phenomena and lead to a many-faceted interpretation of them. Models for an analysis of the general specificity of life in polar areas of our world were furnished by the early investigators of the Arctic – A. F. Middendorff, A. G. von Schrenk, F. R. Kjellman and A. Nordenskiöld.

The tradition of interdisciplinary studies to synthesize aspects of Nature is very strong among Soviet tundra scientists. Being devoted to the tundra, B. N. Gorodkov was at the same time a geobotanist, a florist and a soil scientist. His works comprise thorough analyses of flora, vegetation cover and soils. Many of these papers, reaching back over several decades, still represent a collection of scientific treasures. Another well known student of the northlands, B. A. Tikhomirov, elucidated such problems as the role of plants in the nourishment of animals and the effect of animals on the vegetation, the dependence of plants and animals on soil structure and the effect of various groups of organisms on the soils, the relief, etc. One of the first biocoenological investigations in our country was carried out in the subarctic areas of the Kol'skiy Poluostrov by Fridolin (1936), more than 40 years ago. Another example of a truly ecological–geographical nature study is the comprehensive work, *The Subarctic*, by Grigor'yev (1956).

In the Far North the necessity for a complex approach to utilization of the landscape and exploitation of natural resources is felt especially strongly. In the Arctic, as nowhere else, life of man depends more obviously on the natural surroundings, the extremes of which are immeasurably more severe than in any other zone. To alter the nature of the Northlands it is necessary to proceed with great care because it is very sensitive; when violated this nature is extremely hard to restore and often it turns against man with unforeseen consequences resulting from his foolish treatment of it. There are numerous problems regarding the utilization of the arctic areas, which have no analogy in other types of landscapes. Specialists working out principles and methods for a rational utilization and protection of the natural resources must study a multitude of data from the widely different spheres of research on the Arctic: climatology, soil science, permafrost studies, botany, zoology, etc. For this it is necessary to have reviews and popular books, illustrating the various aspects of life in the landscapes around the Pole.

In this book I am suggesting a preliminary stage for the development of a moderate management of the tundra landscape based on contemporary tundra science. I have written in the form of a short review of some of the most interesting ecological–geographical problems. During preparation of this book I have received much help from the botanists N. V. Matveyeva and B. A. Yurtsev as well as the zoologist A. V. Krechmar who kindly provided photographs of birds and animals. V. G. Kovalev and V. S. Shishkin have donated some of the drawings. To all these I extend my deep-felt gratitude.

2

What is the tundra?

Tundra is the common term for a treeless landscape situated north of the taiga along the coast of the Arctic Ocean, and its islands. The word 'tundra' is derived from the Finnish word *tunturi* (completely treeless heights). In this sense it was originally applied to Kol'skiy Poluostrov. The word is now met with in the languages of many people, even those living in forested areas. In southern Siberia, Kamchatka and on Sakhalin various types of bogs are sometimes called tundra.

In a physico–geographical sense there is also a 'mountain-tundra', i.e., the treeless landscape of alpine mountain belts. Mountain tundra is especially widespread in Siberia. On mountains at lower latitudes, for instance in the subtropics, a tundra type of vegetation is less characteristic. There, alpine meadows have developed above the upper tree limit.

It is necessary to keep in mind that the word 'tundra' just like 'taiga' and 'steppe' has a double meaning. On the one hand it is used as a shortened form of 'tundra zone', as when speaking of the 'taiga' and meaning the 'taiga zone'; on the other hand it denotes a special type of vegetation association, necessary to distinguish it from, for instance, flood plain vegetation, bogs or patches of forest distributed within the tundra. In an analogous sense there is not only taiga inside the taiga zone but also meadow, bog, flood plain, etc. There is, however, no real reason to take a negative view of the double meaning of this term. Below the word tundra will be applied in both the wide and the narrow sense.

We are, in this book, concerned only with that territory which can be considered as a natural tundra zone according to physico–geographical concepts. This means the territory where, in the lowlands, peaty glei soils have developed, carrying a vegetation in which the main role is played by mosses forming continuous mats or separate patches. Beside the mosses, lichens are also typical of the vegetation. The flowering plants consist

3

mainly of sedges (*Carex*) and cotton-grasses (*Eriophorum*); but also of grasses, bushes such as willows, birches and alders (*Salix, Betula, Alnus*), dwarf shrubs (*Betula nana, Dryas*), various species of Ericaceae including bilberries, whortleberries and cowberries (*Vaccinium*) and a few herbaceous dicotyledons. However, the mosses are the basic 'building blocks' of the associations or, as the ecologists say, the 'environment-forming' plants (just as, for instance, is spruce in a spruce forest). In the tundra they indicate both the summer temperature of the soils and the depth of the thawed zone above the permafrost. Also they serve as habitats for many animals, herbaceous plants and shrubs. The tundra specialists often agree with the brilliant words of Lomonosov in his poem *On Wintry Paths*: 'and they call the tundra places overgrown with moss'.

Fig. 1. The basic zonal landscapes of the Eurasian Arctic. 1, The polar desert; 2, the arctic tundra subzone; 3, the typical tundra subzone; 4, the southern (shrub and tussock) tundra subzone; 5 southern limit of the forest tundra ecotone.

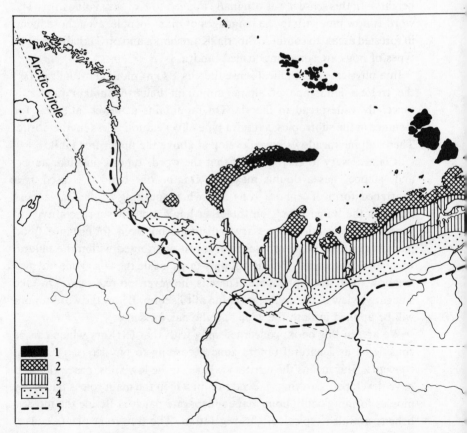

The mosses reach from the tundra into the taiga where they also form a continuous cover. In the literature we very often meet with the opinion that the tundra is a taiga without trees. The reason for this is a certain resemblance between their climatic conditions (e.g., much moisture, long and cold winters, etc.) and the common history of these territories. The number of plants and animals common to both tundra and taiga support these ideas.

The Eurasian tundra zone extends along the coast of the Arctic Ocean (see Fig. 1). This does not, of course, mean that the tundra is a coastal type of landscape. The fact is, that the coastal area along this ocean falls more or less within the latitudes of the subarctic belt where mainly tundra has developed. At the same time the sea strongly affects the tundra landscape by displacing its borders and causing changes to its vegetation, soils and animal life.

The general area of the lowland tundra is definitely not small. It covers about 4×10^6 km², but this amounts to only a small portion of the entire

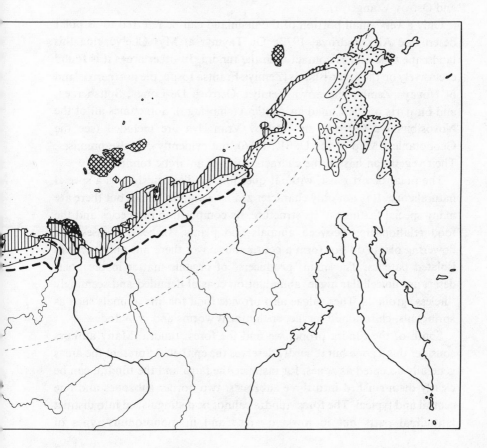

Eurasian continent, approximately 3 per cent. This is a result of the fact that a major part of the tundra has been 'drawn from the sea'. This is, however, not the only reason. Its extension to high latitudes is comparatively limited. The length of the equator is 40076 km, but that of the 70th parallel amounts to only 13752 km. More than half of the tundra zone is located within the borders of the USSR.

The width of the Eurasian tundra belt varies in different areas. Thus, on Taymyr it extends from 600 to 700 km from south to north, in the south bordering on the forest tundra and in the north on the polar desert. However, in other places the continental subarctic landscapes are sharply limited by the coastline. Thus, west of the Yugorskiy Poluostrov only a narrow belt of the southern type of tundra has developed. The same is the case on Kol'skiy Poluostrov, and in Scandinavia it is completely lacking. There we meet only with mountain tundra. East of the Taymyr lowland, too, the tundra extends along the coast as a narrow belt. Its northern variant is there represented mainly on the islands (Novosibirskiye Ostrova and Ostrov Vrangelya).

Only a very small portion of the mainland can be referred to as polar desert (see Aleksandrova, 1977). On Taymyr at Mys Chelyuskina this landscape type makes contact with the tundra. In other areas it is found exclusively on the arctic islands (Zemlya Frantsa Iosifa, the northern island of Novaya Zemlya, Severnya Zemlya, Ostrova De-Longa, Spitsbergen, and on parts of the Canadian Arctic Archipelago). Sometimes all of the Novosibirskiye Ostrova and Ostrov Vrangelya are included (see the Geobotanical Map of the USSR, 1956), but evidently on false premises. Their vegetation has all the characteristics of an arctic tundra.

The polar desert must, without question, be distinguished as a special natural zone. It is not only characterized by a scarcity of life but there are many special features in its structure, its composition of species and the food relationships between animals and plants. Lichens, mosses and flowering plants do not form a continuous cover there but only occur in isolated patches. The main 'producers' of organic matter in the polar desert are unicellular algae, abundant on coastal denuded and seemingly 'lifeless' grounds. These algae also provide food for tiny animals such as springtails, chironomid larvae, enchytraeid worms and nematodes.

South of the tundra proper we find the forest tundra. Many authors consider this a zone but it hardly deserves the epithet. At present the areas generally accepted as zones, for instance the taiga and the tundra, can be clearly distinguished into three subzones, two border subzones and one central and typical. The forest tundra cannot be distinguished into distinct latitudinal parts but in lowland areas and in monotonous types of landscapes it is roughly equivalent to either a tundra or a forest subzone. It could actually be called a transition belt between the tundra and the taiga [cf. forest–tundra ecotone, see translator's foreword] or a zone of secondary order (Chernov, 1975). Thus, the two types of forest zones, the taiga and the deciduous forest, are separated in the north from that other pair of zones, the tundra and the polar desert, by the forest tundra. In an analogous fashion there is also a transition belt between the forests and the arid territory, that is, the forest steppe. Such landscapes, because of their intermediary position, play an important role in the formation of the flora and the fauna and serve as sources of species associations for adjacent zones. This development results in a so-called border effect which extends the variety of the living species and intensifies the processes at the boundary between different types of environment (just as, for example, between a body of water and dry land). Great importance has been placed on the role of the forest tundra for the development of the arctic flora (Yurtsev, 1966).

The entire treeless territory around the North Pole is often called the 'Arctic'. In a narrow sense as used by specialists, this means the basin of the Arctic Ocean and its ice-covered islands, the polar desert and the variant of the tundra occurring at the highest latitudes. The territory situated further south, i.e., where the tundra landscape has mainly developed, is called the 'Subarctic'. Grigor'yev (1956) used these terms to designate types of physico–geographical environments, including in them mainly the geophysical characteristics. As a territory, the subarctic does not fully agree with the extension of the tundra zone. Grigor'yev included the forest tundra in the subarctic. According to S. V. Kalesnik's *Encyclopedia of Geographical Terms* the northern limit of the subarctic belt runs along the mean July isotherm of $+5$ °C and its southern border along the $+12$ °C isotherm. At the northern border of the tundra zone (in the narrow sense) the July temperature varies from 1.5 to 3 °C but at its southern limit it is somewhat below 12 °C (locally 10 °C). Thus by comparison with the tundra zone (in the strict sense) the limit of the subarctic (in the strict sense) runs somewhat further south. The subarctic (in the strict sense) does not reach as far north as the tundra (in the strict sense), but in the south it includes the forest tundra. In this context the subarctic can be considered the same as the 'low-arctic', i.e., part of the Arctic. What lies north of the subarctic is then the true Arctic or the 'high-Arctic'. In this text the term 'arctic' will be used both in the wide and the narrow sense, that is, as a designation of all the entirely treeless territory around the North Pole as well as the basin of the Arctic Ocean and its islands on which ice cover and polar deserts have developed.

By 'subarctic' non-Soviet authors often mean areas which can definitely be considered as 'south of the Arctic' in the wide sense of that term. Thus, some American scientists have given the epithet 'subarctic territory' to an area located south of the northern forest limit and they include in it the forest tundra as well as the northern and middle taiga subzones. This is hardly correct. The use of the word 'arctic' (in any combination) as a name for adjacent forested territory leads to terminological confusion (see Löve, 1970, cited in the Translator's Preface). There must be a priority for what has truly, according to Russian scientists, been included under those terms and based on their physico–geographical knowledge of arctic regions.

The term 'subarctic' is most sensibly used in the meaning applied by Grigor'yev. From the point of view of the utilization of arctic lands and near-arctic regions it is always possible to apply the epithets: 'northland', 'far north' to 'polar lands' which do not have any physico–geographical connotations.

3

Temperature and humidity
in the tundra

Natural zonation is, first and foremost, the result of an uneven distribution of solar heat. Various kinds of solar radiation can be distinguished: direct, scattered, total amount, and so on. 'Direct solar radiation' is when the heat waves reach a place directly exposed to the sun's rays. Its extent may diminish depending on the transparency of the atmosphere and the elevation above sea level but is, of course, also dependent on the latitude. At high latitudes the direct radiation is minimal. When reaching the atmosphere, more or less of the direct rays are scattered as a result of clouds or moisture in the air, but some always penetrate to the surface of the ground. This is 'scattered radiation'. In polar regions the amount of scattered radiation is, to a considerable extent, due to the high level of moisture present in the air.

To characterize temperature, 'total solar radiation' and the so-called 'radiation balance' are of importance (see Table 1). Total solar radiation means the combination of direct and scattered radiation. It can vary according to the time of day, month, season or year. Its magnitude varies

Table 1. Mean annual radiation indices within the different zones along a line from Dikson to Kyzyl, kcal cm^{-2} (Source: *References to the climate of the USSR*. Leningrad, 1967.)

Observation area	Total radiation	Radiation balance	Effective radiation	Albedo %
Tundra	67	16	18	50
Northern taiga	79	25	24	38
Southern taiga	87	32	25	35
Steppe	119	49	34	30

from 60 kcal cm^{-2} yr^{-1} in the Arctic to 200 kcal cm^{-2} yr^{-1} in the Tropics. In lowland areas its isopleths follow in general along the lines of latitude.

The main indicator of the heat condition is the radiation balance at the ground surface or the so-called 'residual radiation', that is the relationship between the total radiation reaching the ground surface and the part radiating back into space. In other words, this is the quantity which can be utilized for the different processes taking place in the ground cover. The yearly amount of radiation balance ranges from 100 kcal cm^{-2} yr^{-1} in the Tropics to a few units or even negative amounts at the Poles. The radiation balance depends on the capacity of the ground surface for reflection, the so-called 'albedo'. This is of course especially great in the polar areas which are covered by snow and ice during most of the year. The albedo of a snow cover can reach 95 per cent.

Total radiation in the tundra region amounts to around 70 kcal cm^{-2} yr^{-1}. The isopleth of this amount in Eurasia runs mainly across the middle of the tundra zone, bending southward in the west, in areas near the Atlantic, and deviating northward further east. At the

Fig. 2. Seasonal variation of solar radiation intensity in the Bol'shezemel'skaya tundra (Grigor'yev 1956); 1, total radiation; 2, radiation balance of exposed surface.

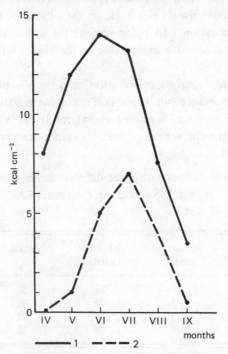

northern border of the tundra, the amount of total radiation is about 64 kcal cm^{-2} yr^{-1}.

The 'isolines' of the radiation balance agree better than lines of latitude with the boundaries between the zones and the subzones. Within the tundra zone this balance varies from about 10 to 20 kcal cm^{-2} yr^{-1}. It averages 15 kcal here and this amounts approximately to 21 per cent of the total radiation. In the polar desert it stands for about 16 per cent. Along the more southerly latitudes, for example in the forest belt, on the steppes and in the southern desert zones, the radiation balance amounts to 30–35 per cent of the total radiation. This is primarily the result of the high rate of albedo in the polar landscapes where snow cover is present for the major part of the year.

The seasonal temperature pattern in the tundra zone is very peculiar (see Fig. 2). This also leads to the typical characteristics of the ecological conditions in the subarctic. From November to February – the time of the arctic night – the mean monthly total radiation in the tundra is about zero. The general amount of winter radiation is negligible (see Table 2). In March it increases to 3–4 kcal cm^{-2} yr^{-1} but almost all of it is lost to the back-scattering effect and the albedo. During the lengthening arctic day in April the total radiation increases to 8–9 kcal and in May to 12 kcal. But also at this time the snow cover persists, reflecting the major part of the radiation so that the radiation balance at the ground surface under it remains very low (0 kcal in April but 1–3 kcal in May).

The weather during April remains winterlike over the major part of the subarctic with hard frosts and snowstorms. Hard frosts persist into May as well and blizzards will occur but the temperature begins to oscillate strongly, sometimes approaching thawing. About the middle or end of May signs of spring begin to show in the southern parts of the tundra zone:

Table 2. Comparison between the magnitude of annual and total summer radiation in the eastern European subarctic (according to Grigor'yev, 1956)

Observation area	Total annual radiation kcal cm^{-2} yr^{-1}	Total radiation April–Sept. kcal cm^{-2}	%	Total radiation June-Aug. kcal cm^{-2}	%
Northern part of tundra zone	64	58	91	35	55
Southern part of tundra zone	70	61	87	36	51

bare patches of ground appear, clear days alternate with warm, cloudy days and days when wet snow falls. However, persistent blizzards and low temperatures may still occur through to the middle of June. In July the total radiation has increased to a high of 14 kcal. During this time, however, the albedo drops sharply as a result of the ground being bared and the diminishing reflective capacity of the melting snow. The July radiation balance is thus 5–6 kcal. Snow remains until the end of June or the beginning of July (not taking persistent snowfields into account).

At the end of June the solar radiation starts to increase as a result of the time during which the sun stays above the horizon. The mid-July total radiation is about 13 kcal. The radiation balance increases also during July. It reaches maximum around the middle of this month but falls off sharply during August and September to reach a minimum in October.

What are the main characteristics of the 'temperature pattern' in the subarctic? First of all, the graph representing the radiation balance has a single peak. It rises steeply from May to July but drops sharply during August and September. The period of comfortably warm temperatures in the tundra is very short. This contributes to the extraordinary dynamics of the seasonal development of the living organisms and the brevity of the phenological periods. Secondly, there is a sharp difference in time between the peaks of total radiation and that of the radiation balance. In April and May the quantity of solar radiation in the tundra zone is equal to the monthly mean in subtropical areas but the immense quantity of energy is lost to biological processes because the radiation balance is still so low. At the time when this reaches a state where the solar radiation can effectively be utilized by living organisms, it is already beginning to decline.

To most people living around the North Pole or in the Arctic, tundra means the realm of frost. It is, however, not so much the cold temperatures as such which determine the main peculiarities of the living environment in the arctic areas. The coldest areas in the north are not found in the tundra but inside the taiga. For example, Verkhoyansk – best known as 'the Pole of Cold' – is situated south of the tundra. Summers are, of course, much hotter there than in the Arctic. In addition, in eastern Siberia, below the Arctic Circle and at the southern boundary of the tundra and in some places inside its borders, steppe-like associations are found which display plants and animals characteristically requiring hot and dry summers (Yurtsev, 1974). The summers in those areas, e.g., in northern Yakutia, can actually be very hot: in July the temperature there may reach 30 °C.

What then are the most important factors responsible for the development of landscapes within the tundra zone and the characteristics of its flora and fauna?

The mean yearly amount of heat supply, arriving in the form of solar radiation, depends on the radiation balance at the ground surface. But not all of it is equally important for living organisms. The main role is determined by the nature of the summer temperature. It is exactly the summer temperature which first and foremost reflects the main factor in zonation, that is, the amount of radiation balance. The winters are strongly dependent on the degree of continentality, i.e., the remoteness from a warm ocean not covered by ice. Over most of its area the tundra zone is a coastal land but the Arctic Ocean is covered by ice the major part of the year and therefore, from a climatological point of view, it forms a single unit together with the continent. Here, air masses with very low temperatures are formed. In the west, the ice-free Atlantic constitutes a powerful climatic factor. Its effect on the period of cold temperatures exceeds the influence of the radiation balance.

The maximum temperature in the tundra can be very high. For example, in northern Taymyr the air temperature in July can reach 20 °C and the absolute maximum at latitude 73 °N has been measured at 29 °C. In the southern parts of the subarctic, the maximum air temperature can stay at 25 °C for several days in a row.

Still, the level of the maximum temperature is not the decisive factor for the development of organic life in the tundra. Rather, this depends on the length of the warm period. Some species of animals, mainly birds and mammals, are active under arctic conditions all year round, such as foxes, polar bears, ptarmigan (*Lagopus mutus*) and the northern reindeer. Some, for example, lemmings, even reproduce during winter in the tundra. But the major part of the tundra associations is active only during summer (i.e., the vegetation, micro-organisms, invertebrates). During the summer all the basic abiotic processes operate in the landscapes: erosion, frost heaving, thawing and permafrost phenomena, etc. Therefore the length of the frost-free period is of primary importance for life in the tundra. The main characteristics of the tundra landscape and its organic life depend on it.

The completion of the life cycles of animals and plants depends on the length of an adequately warm period. Below it will be demonstrated that it is just in the life cycles that the importance of the adaptability of animals and plants to tundra conditions is manifest.

A significant detail of the temperature regimen of the Arctic is the very wide variation in temperature. This can easily be seen when we look at the course of the isotherms of the mean monthly temperatures during the warm period in the tundra zone and those at more southerly latitudes (see Fig. 3). Thus, in the Bol'shezemel'skaya Tundra area the mean July temperatures vary from 6 to 12 °C over a distance of 300 km. The July temperatures across all the Russian lowland, stretching 2000 km from the Arctic Circle down to the 50th parallel and including four natural zones, vary by almost the same amount (from 13 to 20 °C). Sharp climatic changes are of course not characteristic of the arctic areas only. This is a characteristic feature of landscapes which are developing in the absence of heat and moisture. In the deserts the variability of the climate is as sharp as that in the tundra.

The wide variation in the climate of the arctic areas is reflected in the structure of the landscape and the organic life of the tundra and determines the wide differences between the subzones.

Fig. 3. Mid-July isotherms (°C) over the territory of the Russian lowland (Mil'kov and Gvozdetskiy 1976).

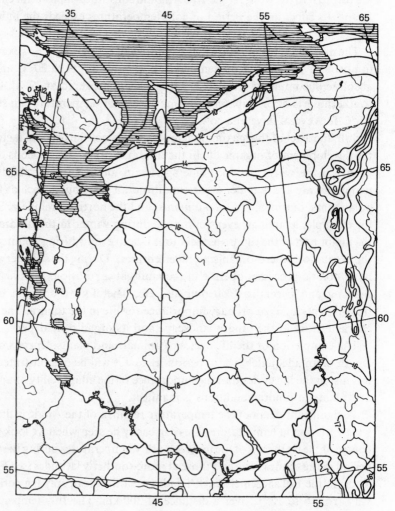

It is a well known fact that there is much water in the tundra and that both soil and air have a high moisture content. Large areas within this zone are occupied by bogs and mires [see translator's foreword]. This is sometimes called the tundra bog-belt. However, this is not the result of heavy precipitation. If moisture were to be defined only in terms of precipitation, the tundra could be considered one of the most arid types of landscape (see Table 3). In various parts of the subarctic precipitation varies from only 150 mm up to 350 mm per year, i.e., about as much as in semi-deserts, dry steppes and savannahs. Because of this, the polar deserts, situated north of the tundra, have derived their name since no more precipitation falls there than in the hot deserts, that is, less than 200 mm per year. Nevertheless, in the tundra as well as in the polar desert there is always a high degree of moisture.

The abundance of moisture is a direct result of *a low rate of evaporation* which in turn depends on insufficient heat. Besides precipitation in the form of rain and snow in the subarctic there is also another important source of moisture, i.e., the accumulation of hoarfrost from the saturated atmosphere on the cool surface of the snow. This occurs especially often during invasion of warmer masses of air (Grigor'yev 1956).

The tundra is wetter than any other landscape on Earth. Only some parts of the areas of boggy taiga, e.g., in western Siberia, can rival it with respect to abundance of water. The landscape-forming role of water is nowhere as distinct as in the tundra. From an aeroplane it has an aspect of an almost continuous lacework of innumerable lakes and ponds formed as a result of melting permafrost and subterranean ice. There are many bogs and mires in the tundra. Their characteristic structure of polygons, water-filled cracks and peaty hummocks, are formed as a result of cryogenic processes.

Table 3. Humidity conditions at various latitudes (according to Kalesnik, 1970; Ryabchikov, 1972)

Zonal type of landscape	Amount of precipitation yr^{-1} (mm)	Radiation index of dryness
Polar desert	110	0·1
Arctic tundra	200	0·2
Typical tundra	250	0·3
Taiga	370	0·6
Mixed forest	450	0·8
Forest steppe	380	1·0
Desert	100	3·0
Tropical forest	1000	0·9

Subterranean ice, water melted from the snow, fog and long-lasting rains mixed with ice or snow, are all powerful ecological and landscape forming factors in the tundra.

Temperature and humidity are closely interdependent within a landscape. The higher the temperature, the more water that can be evaporated. Evaporation leads, however, to dispersal of heat. Consequently, the more water evaporated, the stronger the cooling effect. Atmospheric humidity is a result not only of the amount of precipitation but also of the amount of incident solar heat. This can be high during very heavy as well as during very low precipitation. Therefore, when characterizing a climate, not only the amount of heat and atmospheric humidity but also their correlation must be considered and this has been called the 'hydrothermal index'. The widely known 'radiation aridity index' of Budyko (1948, 1950) is the ratio between radiation balance and total yearly precipitation calculated on the basis of latent heat of evaporation.

This index indicates in equal units the proportion of heat and atmospheric humidity where the amount of evaporation corresponds mainly to the amount of precipitation that has fallen. Territories with a fair balance between temperature and precipitation are more favourable for the development of organic life. No excessive rainfall or severe droughts occur in them and as a rule it is adequately warm (e.g., in areas of broad-leaved forests and forest steppes).

In deserts belonging to the temperate belt this index increases to 3–5 but in tropical arid landscapes to 10. North of the forest steppe and the deciduous forest, the index is expressed by smaller numbers: in the taiga it amounts to about 0.6, in the tundra to 0.3, in the polar desert to 0.2 or less. In these zones conditions develop for an accumulation of surplus moisture: the temperature is not high enough for evaporating the yearly amount of precipitation. The territories where zones of mixed forests, taiga, tundra and polar desert occur are therefore called the humid belt (from the Latin word *humidus* for moisture).

An index of about 1 is typical not only for landscapes of the temperate belt but also for the subtropical and equatorial rain forest. The ratio of 0.3 is characteristic of tundra as well as of some subtropical and tropical territories with a predominance of boggy forests. This recurrence of a balance between heat and atmospheric humidity in various climatic belts has been described in papers by Budyko and Grigor'yev under the term of 'the periodic law of geographic zonation'. According to this law the tundra possesses some traits in common with those of strongly humid landscapes in the subtropics and tropics (e.g., some elements of soil-forming processes). A similar relationship between heat and humidity in these –

admittedly – not very similar landscapes determines undoubtedly the similarity of some of the characteristics of the living beings. This emphasizes the uniformity of the landscapes covering the Earth and interrelates them with as well as subjects them to general laws, even when they occur under diametrically opposite climatic conditions.

The lack of heat is the decisive factor causing the extremes of living conditions in the Arctic. These are different from those of the hot deserts where there is much heat but little moisture. The fact is that water exists in some form also in the deserts (partly in inorganic form) although only distributed patchily. In order to survive in a desert it is necessary to be able to extract the water, to store it and conserve it for prolonged periods of time. Therefore, in the process of evolution, desert organisms have, no doubt, developed different and at times astonishingly effective adaptations for guaranteeing a supply of water (deep root systems, deep burrows, storage tissues, minimal rate of evaporation, etc.). Life in a desert cannot exist everywhere, it is localized and it is also very variable.

The lack of heat in the tundra is practically impossible to counteract by means of any kind of adaptation. Here only the most general characteristics of major groups of organisms may be of significance, e.g., warmbloodedness (homoiothermia) or coldbloodedness (poikilothermia). Partial adaptations directed against cold are also limited and cannot simultaneously be the backbone of variations for adaptations to drought. It is, however, significant that on the tundra there are also many plant characteristics clearly adapted to the conditions of water supply, about which more will be said later. Such morphological peculiarities of animals as increased body surface, increased heat insulation and temperature protective fur or the biotopic distribution of species (i.e., the selection of warm habitats on south-facing slopes etc.) are of considerably lesser importance.

Heat is the main condition for the operation of all biological processes; lack of it makes life in any form impossible. Therefore the organic life in the tundra zone is limited quantitatively (in number of species and types of adaptations) but instead it is spread over the landscape more uniformly than life in a desert.

4

The diversity of tundra landscapes

The tundra zone is far from uniform in its different parts. It is shaped by the latitudinal–zonation characteristics of the climate (i.e., divided into subzones), the degree of continentality and the variation in relief. Below we will discuss the factors within its geographical area which form the tundra landscape.

Subzones

Above we have mentioned that the temperature of the warm season is the most important factor for organic life in the subarctic. It diminishes much faster from south toward north than within the temperate belt. At the same time, where there is a general lack of heat, any rise in temperature is strongly reflected in the life processes of the organisms and in the structure of the landscape. This is expressed in the sharp latitudinal differences between the subarctic landscapes and by the characteristics distinguishing the tundra zone into subzones. Usually three subzones are distinguished: the southern (or shrubby) tundra, the typical tundra and the arctic tundra (see Fig. 1).

In the 'southern tundra subzone' a tier of bushes about 0.5 m high, rarely reaching 2 m, has developed on the main surface of the watersheds [see translator's foreword] (see Fig. 4). This tier consists locally of birches, willows and alder (*Betula*, *Salix*, *Alnus*). Herbaceous plants, sedges (*Carex*), cotton-grass (*Eriophorum*) and grasses, shrubs and dwarf shrubs such as dwarf willow (*Salix herbacea*), bog whortleberries (*Vaccinium uliginosum*), cowberries (*V. vitis-idaea*), Labrador tea (*Ledum* spp.), usually grow in the shelter of the taller bushes. Below them a continuous cover of bryophytes has formed, consisting of many species. The most common are *Hylocomium splendens* var. *alaskanum*, *Aulacomnium turgidum*, *Toment-hypnum nitens*, *Rhacomitrium lanuginosum*, *Ptilidium ciliare*, and species of *Dicranum* and *Polytrichum*.

18

In the southern tundra subzone individual trees are met with, almost all belonging to the larches (*Larix*). They are low-growing, have contorted, slender stems or assume a krummholz habitat (see p. 57). A few thick half-rotten stumps, suggesting the idea of a cooling climate, is all that remains of forests once advancing from the south or of the intensive attempts of man to cultivate trees. The height of the bushes in the tundra indicates the thickness of the snow cover: jutting up above the snow are some slender branches, during winter subjected to so-called snow abrasion (the snow melts away around the branches, the wind abrades them). This, however, does not occur everywhere and not to all species of bushes. Alder can, for instance, reach up to 2–2.5 m in height on slopes, which certainly exceeds the snow cover in those habitats.

The main indicator of the severity of the arctic climate in this subzone is the lack of real tree vegetation but otherwise the southern tundra is characterized by very rich associations. There is always a very variable flora and fauna. Beside species typical of the tundra there are representatives also of middle latitudes. For instance, in the European and Siberian southern tundras it is possible to find many plants common to the temperate belt such as marsh cinquefoil (*Potentilla palustris* = *Comarum palustre*), golden saxifrage (*Chrysosplenium alternifolium*), marsh marigold (*Caltha palustris*) and even the heat-loving thyme (*Thymus serpyllum*). Among the birds we may encounter examples of the willow warbler

Fig. 4. Landscape in the southern tundra subzone in Taymyr.

(*Phylloscopus trochilus*), redwing (*Turdus iliacus*), snipe (*Gallinago gallinago*) and short-eared owl (*Asio flammeus*). Around the lakes, pin-tails (*Anas acuta*) nest and of the tundra rodents, the root vole (*Microtus oeconomus*) has a wide distribution.

A very variable vegetation cover has formed in the southern tundra. On watersheds thickets of willows, birches (including dwarf birches) and alder alternate with shrubless tundra having an unbroken moss cover or patches of barren ground. In hollows, various bogs or mires of *Hypnum* or *Sphagnum* type have formed being either flat or with peaty hummocks. On south-facing slopes, where much snow accumulates, a vegetation cover of grasses, species of the Leguminosae and various herbs has developed. On elevated rims a dense growth of dwarf shrubs occurs: cowberries (*Vaccinium vitis-idaea*), bog whortleberries (*Vaccinium ulginosum*), crowberries (*Empetrum nigrum, E. hermaphroditum*) and black bearberries (*Arctostaphylos alpina = Arctous alpina*), etc. Near water, around lakes and on the banks of rivers various types of wetland vegetation occur with species of sedge (*Carex*), mare's tail (*Hippuris*) and grasses among which *Arctophila fulva* is especially characteristic.

Such a mixed vegetation is very typical of the forest tundra. The combination of different typical associations on zonal mesic habitats [plakors, see translator's foreword] is a general characteristic of a transitional zonal category, an ecotone. Let us think of forest steppes and semi-deserts: on watersheds in the forest steppe, both steppe and forest associations alternate and in the semi-desert there are both steppe and desert coenoses. Shrubs and even trees thrive there. The southern tundra landscape of Taymyr, with its thickets of alder, is strikingly reminiscent of the cis-Caspian semi-deserts where among the sagebrush (*Artemisia*) vegetation a tier of meadow-sweet (*Spiraea*) has developed.

In the subarctic parts of eastern Eurasia, especially on the Chukotka peninsula, the shrub associations in the southern subzone are replaced by a so-called tussocky tundra where cotton-grass (*Eriophorum vaginatum*) forms thick, perennial tussocks among which birch and willow grow thinly.

The southern tundra is in many respects similar to the forest tundra. These two have many species of plants and animals in common. Special groups of 'hyparctic' species can be picked out, which are characteristic of the areas both of southern tundra and of forest tundra. Such are Middendorff's vole (*Microtus middendorffi*), the willow grouse (*Lagopus l. lagopus*), the bar-tailed godwit (*Limosa lapponica*) and the spotted redshank (*Tringa erythropus*) and the carabid beetle (*Carabus odoratus*) as well as the largest of the tundra spiders, *Lycosa hirta*, and among plants the dwarf birch (*Betula nana*). They have also similar edaphic associations.

In the southern tundra just as in the forest tundra various bushes (of birch, alder and willow) form closed thickets, their roots piercing the peaty soils and in autumn their leaves falling in a thick layer. This is of great importance for the existence of herbaceous species and animals. In the southern tundra, for instance, the typical inhabitants of forest litter, the colony-forming larvae of the gnats (*Bibionidae*), are still common, but they are absent from the typical tundra where there is very little leaf litter.

Although it may be expedient, as is done by some authors when distinguishing landscapes, to separate the southern tundra and the forest tundra from the rest of the subarctic, there is no doubt that with respect to many important characteristics in the structure of these landscapes, the vegetation cover and the animal inhabitants, all the area north of the forest tundra is characterized by many common traits which make us combine it in a single zone, the subarctic.

The 'typical tundra subzone' is situated north of the southern tundra. Its boundaries run mainly along the July isotherms for 8–11 °C in the south and 4–5 °C in the north. The potential area within this subzone is larger than that of the other subzones. On Taymyr it has a width of 300–400 km while the width of the arctic tundra subzone, there, is 200 km and that of the southern tundra only 150–200 km. However, in general, the coast line along the Arctic Ocean 'cuts off' a major part of the continent where this zone would otherwise have existed. In Eurasia it is well represented on Taymyr, Yamal, Gydanskiy Poluostrov and Yugorskiy Poluostrov and between the Yana and Kolyma rivers, but the rest of it consists of smaller areas only and in the south of merely some fragments. It is completely lacking on the continent west of the Yugorskiy Poluostrov (see Fig. 1).

This subzone is the most characteristic of the landscapes we call tundra. Here, not only trees but also tall bushes are absent on the watersheds. The height of the vegetation is entirely determined by the thickness of the snow cover. As a result of snow abrasion during winter, only those plants remaining snow-covered can survive and the thickness of snow is not very great, only about 20–40 cm. Shrubby thickets may reach a height of 1 m in depressions, in river valleys and along lake shores, where more snow accumulates.

The typical tundra is the realm of bryophytes. The thick carpet of mosses, an unbroken layer covering the soil, is the most important part of its vegetation cover. This measures from 5 to 7 cm, locally up to 12 cm, in thickness and consists of many species of mosses among which about 10 species are the most predominant. Three of these are especially typical: *Tomenthypnum nitens, Hylocomium splendens* var. *alaskanum* and *Aulacomnium turgidum*. Also common are *Rhacomitrium lanuginosum* and the liverwort *Ptilidium ciliare* as well as some other mosses, e.g., species of *Dicranum* and *Polytrichum* which form dense hummocks and cushions. Almost all the mosses are also widely distributed in more southerly zones, especially in the taiga.

The moss cover plays a very important but contradictory role in the life of the tundra. Only mosses provide a completely closed vegetation in the watershed areas within this subzone. They have a very important effect on the soil temperature and moisture, and on the dynamics of the seasonal thawing of the ground. On the one hand the moss cover keeps the permafrost from thawing but at the same time it has a negative effect on the development of other organisms. The thicker, denser and drier it is, the cooler is the soil and the nearer to the surface the permafrost will be. In fissures and depressions where there is a specially thick moss cushion, the ground below it often does not thaw until the end of the summer. On the other hand, the moss cover prevents thermokarst (resulting from deeply thawed ground and subterranean ice) from developing and at the same time it helps stabilize the vegetation. The destructive consequences of stripping off the moss cover, e.g., as the result of traffic with tracked vehicles, are well known.

During summer the moss cushion prevents the upper soil horizon from drying out. This is very important when considering the small amounts of precipitation. During spring and autumn it acts as a sponge absorbing water and creating almost bog-like conditions. Stems and roots of rhizomatous species and of herbs penetrate the moss cover. Thanks to the moss layer many species with long roots prosper in the tundra. It is characteristic of plants in the moss tundra that the roots are strongly branched in a horizontal plane within the moss cushion. During the autumn and winter seasons the moss cushion acts as a repository for seeds, but during summer the sprouting plants often do not succeed in taking root in the soil below and perish as a result of drought and lack of nourishment because they are, so to say 'suspended' within the moss cover.

The moss cushion serves as a habitat for a rich complex of invertebrates, called semi-edaphores (i.e., partly living in soil). To these belong the majority of collembolans, mites, spiders and staphylinid beetles. Living together with them in the moss cover are typical soil inhabitants such as earthworms, enchytraeids, larvae of tipulids, carabid beetles, etc. The lemming, most abundant of the tundra animals, owes its life to the moss cover. It digs its entire labyrinth of tunnels within the moss sward and during the winter it lives on the fleshy parts of flowering plants, buried in the cushion. Because they burrow in it, these animals are forced to gnaw through enormous quantities of moss.

The herbaceous tier is formed mainly of sedges among which *Carex ensifolia* is abundant. It is the most common of all flowering plants in the Eurasian tundra, where it occurs usually in the form of the subspecies *arctisibirica*. The sedges are the second most important plants in the typical tundra because of their role as association components. Due to this the basic zonal association of this subzone is called sedge–moss tundra (in the geobotanical terminology used below they hold the place of the main association formers). In addition there are within the typical tundra many-stemmed bushy willows, shrubs and dwarf shrubs (e.g., dwarf species of *Salix*, species of *Dryas*, *Cassiope*, *Vaccinium*, etc.). Often cotton-grass (*Eriophorum*) and dicotyledonous herbs such as *Saxifraga*, *Pyrola* and Compositae are common. Locally in the moss cushion there are many lichens of various kinds. Depending on the abundance of these plants, the tundra type can be distinguished into willow–sedge–moss tundra, *Dryas*–sedge–moss tundra or lichen–moss tundra, etc.

In Soviet literature the associations belonging to this subzone are often called moss–lichen tundras. This term is, in spite of its wide use, not very suitable because although the lichens are almost always numerous in the tundra, they are not conspicuous from the point of view of forming associations. They are usually a significant admixture of the moss sward only. In the typical tundra subzone the lichens are very variable. Some areas have up to 200 species displaying very different life forms (being foliose, tubiform, fruticose, crustaceous, etc.). Associations where lichens occur as first rank dominants are met with only locally on sandy and stony ground.

The landscape within the typical tundra subzone is quite uniform (see Fig. 5): an even, gently rolling relief with stable vegetation cover on the watersheds. This monotony is broken by river valleys, steep shore-lines, boggy depressions and lake basins.

Beside the basic zonal associations with a closed moss cover within this subzone, a so-called 'spotted tundra' [see translator's foreword] has also developed (see Fig. 6). Patches of bare soil (usually 40–50 cm in diameter) are surrounded by slightly raised borders of a dense moss cushion. The borders of adjacent patches are separated by fissures or hollows, filled with peat or a loose moss sward. Spotted tundra occurs usually on raised terraces. Its formation is related to processes of soil cracking, disruption of the moss cover, 'frost boils' in bare soil and breaks in the surface of wet ground, etc.

Bare patches in the spotted tundra will gradually become recovered with vegetation. In some parts, completely bare patches may be found together with patches almost completely recovered with mosses and flowering

Fig. 5. Landscapes in the typical tundra subzone, western Taymyr.
a, Rolling relief; b, habitat with hummocky sedge–moss tundra in a
locality with patchy growth; c, surface of willow–sedge–moss tundra;
d, spots of bare soil in the processes of colonization.

plants. All this creates very colourful ecological conditions due to the fact that in the spotted tundra, plant and animal life is very variable.

Within the typical tundra subzone it is still possible to encounter animals and plants of extra-arctic origin. However, relatively few of those originating from the south are able to lead a normal life here. It is true that these species are generally widely distributed, inhabiting several zones. Some are indifferent to the conditions of the landscape and are called 'eurybionts'. Such are stoats, crows, seven-spotted ladybirds and various species of moss. Others are related to specific biotopes, e.g., shoreline habitats: marsh marigolds (*Caltha palustris*) and the wagtail (*Motacilla alba*). Among the many species, common to both the southern tundra subzone and the forest tundra, some are absent or very rare in the typical tundra. For instance, here we do not meet with such characteristic inhabitants of the forest tundra and the southern tundra landscapes as the lesser white-fronted goose (*Anser erythropus*), the bar-tailed godwit (*Limosa lapponica*) or the pin-tailed snipe (*Gallinago stenura*). Of small animals, mainly the lemmings remain, but those which are abundant in the southern tundra such as the narrow-skulled vole (*Microtus gregalis*) and Middendorff's vole (*M. middendorffii*) as well as the root vole (*M. oeconomus*) and also the arctic shrew (*Sorex arcticus*) are lacking or rare. The abundant berry-bearing bushes of the southern tundra and the dwarf shrubs such as cloudberries

Fig. 6. General view of sedge–moss spotty tundra in the subzone of the typical tundra. 1, Patch of bare ground; 2, raised border; 3, depression.

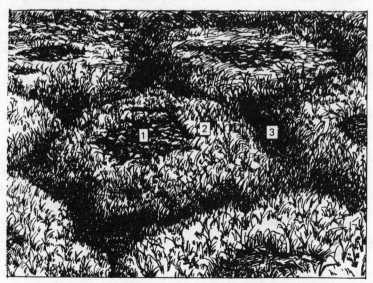

(*Rubus chamaemorus*), bog whortleberries (*Vaccinium uliginosum*) and cowberries (*V. vitis-idaea*) are still found here but do not set fruit as abundantly. In the most northerly parts of the subzone, they are absent.

In addition to the basic zonal sedge–moss tundra described above some quite variable associations have developed in local biotopes within this subzone: there are thickets of bushes along river banks, etc. On the whole the climatic conditions in the typical tundra display only moderate extremes as a result of which the depauperation of plants and animals is not as great as farther north.

The 'arctic tundra subzone' has developed on Yamal'skiy Poluostrov, the Gydanskiy Poluostrov, and Taymyr and, locally, along the shores of the Anabar and the Kolyma rivers, but also on Vaygach, Novaya Zemlya, the Belyy Ostrov, Novosibirskiye Ostrova and Ostrov Vrangelya. The name of this subzone is, perhaps, not the most appropriate because it is often used as a term for the entire tundra area of the Northern hemisphere. However, it has become generally accepted and there is no real reason to discuss it further. This subzone is situated in an area with a distinctly arctic climate. At its southern border the mean July temperature reaches 4–5 °C, at its northern limit only about 1.5 °C. At any time during the summer season, the temperature can fall below 0 °C and snow can fall. The thickness of the snow cover is insignificant because of which the winter conditions for animals and plants are particularly severe.

The main characteristic of the landscape in the arctic tundra is the wide distribution of bare ground. On the watersheds different varieties of associations have developed in which patches of bare ground are surrounded by a vegetation sward. This is called the 'spotted', 'medallion' or 'polygonal' tundra or many other names (see Fig 7). The bare ground occupies about 50% of the surface. The moss carpet with its twigs of dwarf willow and some saxifrages and grasses spreads over frost cracks around the bare spots. Specifically, this type of tundra is reminiscent of the spotted tundra in the southern subzone. But there the bare ground was a secondary phenomenon; it was formed mainly by frost heaving disrupting the moss cover. In the arctic tundra the bare ground is the primary phenomenon and the moss cover is, in general, secondary. The mosses invade the frost cracks which gradually widen as a result of crumbling and settling edges. The arctic tundra is always varied: it is gravelly, stony, clayey or of a genuinely patchy type with a vegetation cover in the form of patches, stripes or networks, etc. The frozen appearance of the arctic tundra subzone is various but noticeable everywhere.

The low degree of decay and the intense cryogenic (frost) action always creates a variable and sharply broken micro- and macro-relief in the arctic tundra. Everywhere there are split blocks and stones. The ground surface is covered with cracks, fissures and mounds. In spite of the fact that the climate is considerably more severe here than in the typical tundra, the landscape still appears more varied and colourful because of the abundance of flowering plants. The conditions existing in this tundra for some groups of organisms are more varied and more favourable than in the southern

Fig. 7. Arctic polygonal, spotted tundra, eastern Taymyr (photo.: N. V. Matveyeva).

tundra where the moss cover creates an unbroken, uniform and monotonous environment. The ground, not protected by moss cover, warms up considerably better in spite of the low temperatures. In the typical tundra subzone on Taymyr, at a mean July temperature of 10.5 °C, the depth of thawing permafrost is about 48 cm but in the arctic tundra subzone it is 50 cm at a mean July temperature of 4 °C.

Therefore the soil of the arctic tundra, although thin, is inhabited by animals in no fewer numbers than that of the southern subzones. The species of soil animals are fewer but locally each species occurs in greater numbers.

At first glance the bare ground of the arctic tundra may appear to be lifeless but in it a rich life of organisms has developed. In the top soil layer, there is an abundance of unicellular algae and nematodes, which feed on them, as well as annular worms of the enchytraeids, collembolans and many larger animals such as earthworms and tipulid larvae. On the surface many crustaceous lichens also carry species of moulds. Flowering plants are distributed among the stones: grasses and rushes (*Juncus*), poppies (*Papaver*), arctic avens (*Novosieversia*), *Dryas*, louseworts (*Pedicularis*), saxifrages (*Saxifraga*), forget-me-nots (*Myosotis*) and many others (see Fig. 8). Some of these form dense cushions or tussocks which locally serve for the accumulation of organic matter during the soil-forming processes.

Fig. 8. Arctic herb–grass tundra with frost cracks, eastern Taymyr.

The plants on the bare ground display interesting morphological adaptations. Thus a saxifrage (*S. platysepala*) propagates by means of thick runners, 'walking' over the soil surface. Many invertebrates in the arctic tundra have adapted to the habitats on the surface of the frozen ground because of the severe winds: some flies and mosquitos lack or have reduced wings. They crawl quickly but fly badly, sometimes using the rudimentary wings as sails so that they get around quickly on the ground while being blown along by the wind. Small soil animals, beetles and collembolans, are generally represented by large species with a dense darkly coloured cover which is characteristic of inhabitants of surface soils.

There are no shrubs at all in the arctic tundra and a very minimally developed herbaceous tier. Certainly, there are many herbs but they do not form a closed cover and often they grow in stripes. The majority of the plants are only 3–10 cm tall; only a few sedges reach up to 20 cm. In the herbaceous tier sedges and cotton-grasses are few but grasses are often abundant. On Taymyr, for instance, the grass *Alopecurus alpinus* occurs everywhere in the arctic tundra, locally forming a closed sward looking almost like a 'meadow'. This is the main difference between the arctic and the typical tundras. In the latter, there are many species of sedge and cotton-grass (*Carex, Eriophorum*) but few grasses. Other important differences are found in the general multitude of dicotyledonous herbs, e.g., the great importance in the vegetation cover of such genera as saxifrage

Table 4. Role of different groups of flowering plants within the associations, Taymyr

Group of plants	Typical tundra	Arctic tundra	Polar desert
Grasses (Gramineae)	+*	+ + +	+ + +
Sedges (*Carex*)	+ + + +	+	−
Woodrush (*Luzula*)	+	+ + + +	−
Shrub willows (*Salix*)	+ + + +	−	−
Dwarf willows (*Salix*)	+ +	+ + + +	−
Birches (*Betula*)	+ +	−	−
Caryophyllaceae	+	+	+ + +
Papaveraceae	+	+ + +	+
Cruciferae	+	+ + +	+
Saxifragaceae	+	+ + + +	+ + +
Rosaceae	+ + + +	+ +	−
Leguminosae	+ +	+	−
Ericaceae	+ + +	−	−
Scrophulariaceae	+ +	+ +	−
Compositae	+ +	+	−

* Visual evaluation of the role in zonal association according to a four point scale.

(*Saxifraga*), whitlow grass (*Draba*), poppy (*Papaver*) and woodrush (*Luzula*) (see Table 4). In the typical tundra these plants occur in various local biotopes, on slopes, in river valleys, on cliff walls, but they are rare in the mossy areas of the watersheds.

In the arctic tundra the number of shrub species is sharply reduced. They are represented here only by *Dryas* and dwarf willow. But these life forms play an essential role in the vegetation cover. The avens, mainly *Dryas punctata*, cover stony ground with a closed mat or by individual cushions. The dwarf willows, of which *Salix polaris* is especially characteristic, grow in large numbers in the moss sward around the patches of bare ground. Most of this shrub lies buried in the thick moss, only the top leaves and inflorescences can be seen above it.

Mosses are very abundant here, just as in the typical tundra. In the arctic tundra they do not form a continuous cover but they have still an association forming function. On ground with cracks they colonize the edges of fissures. When the cracks widen and their edges crumble and cave in, mosses fill the trenches forming a wall instead of a crack around the bare ground. Various flowering plants take root in the thick moss, especially dwarf willows (*Salix polaris*), penetrating the sward with their branches and roots. The lower part of the sward disintegrates and converts into peaty matter. Under the moss cover a peaty layer forms but the mineral ground is barely heated. Ice lenses are found there almost throughout the summer season.

The main characteristic of the specific composition of the flora and fauna in the arctic tundra is the very high percentage of species typical of other zones as well. Just as in the southern subzone, some plants and animals living here have a wide distribution elsewhere. To these belong many mosses, lichens, algae and soil animals, all of which are generally very widely distributed. We also encounter widely distributed animals, e.g., the white wagtail (*Motacilla alba*), the wheatear (*Oenanthe oenanthe*) and the stoat (*Mustela erminea*). However, the number of such species in the arctic tundra is many times lower than in the typical and southern tundras. On the watersheds, which serve as habitats for the majority of the flowering plants, insects and birds, there are typically arctic species characteristic of this subzone, which only there reach their largest numbers. The following may be cited as examples: the ptarmigan (*Lagopus mutus*), the curlew sandpiper (*Calidris testacea*) and the knot (*C. canutus*), the snowy owl (*Nyctea scandiaca*), the craneflies (*Tipula carinifrons*), the polar willow (*Salix polaris*), the flagellate saxifrage (*Saxifraga platysepala*) and the tiny crucifer *Draba subcapitata*. The species representing groups of small soil animals are also generally characterized by a wide distribution. However, in northern Taymyr almost half of the species of collembolans and hard-shelled mites are typically arctic. It is particularly noteworthy that these species are numerous and are abundant in the most diverse biotopes.

In the arctic tundra no species occur which are typical of taiga, forest tundra or southern tundra. There are, e.g., essentially no such species as dwarf birch (*Betula nana*), crowberries (*Empetrum*), arctic bearberries (*Arctostaphylos alpina = Arctous alpina*), cowberries (*Vaccinium vitis-idaea*), bog whortleberries (*V. uliginosum*) or cloudberries (*Rubus cham-aemorus*), willow grouse (*Lagopus l. lagopus*), spotted redshank (*Tringa erythropus*), bar-tailed godwit (*Limosa lapponica*) or Middendorff's voles (*Microtus middendorffii*). Characteristically there are few or none of the many species of poppy (*Papaver*) inhabiting the typical tundra subzone, or species such as the little stint (*Calidris minuta*), the dunlin (*C. alpina*), or a sedge (*Carex ensifolia*).

All this emphasizes the extraordinary specificity and originality of the climatic regime of this subzone. In order to survive here, special adaptations are necessary which allow existence under such severe conditions.

The wide alteration of climate definitely transforms arctic landscapes and leads to the presence of sharply defined subzones. The degree of continentality also plays a great role and is evident, primarily because of the great amplitude of atmospheric temperature. All the territories belonging within the tundra zone are situated close to the oceans. However, at individual places along this coast, the role of the sea varies. The eastern and western borders of Eurasia and America are similarly affected by the waters of the Atlantic and the Pacific Oceans. The warm Atlantic has a particularly strong effect. Along the central part of the Eurasian coast line, the continent comes in contact with the cold Arctic Ocean, a major part of which is ice covered and can therefore be considered as 'dry' as a result of which the climate here becomes continental.

It should be mentioned that a warm sea is capable of making the climate milder and of displacing the limit of the tundra further north. However, the edges of the continent subjected to the effects of a warm sea have a treeless tundra stretching far southward. Thus along the Pacific Ocean, tundra has developed almost down to the 60th parallel. At this latitude a middle-taiga forest has developed in the centre of the continent. Even on Komandorskiye Ostrova, situated at latitude 55 °N, the landscape has the character of a tundra. In Europe, the southern limit of the tundra is at latitude 66.5 °N while on Taymyr it only begins at latitude 72 °N. This is determined first and foremost by the peculiarities of the seasonal course of the warming processes under continental and oceanic conditions of climate. In areas with a high degree of continentality, there is less cloudiness therefore temperature maxima are reached sooner after the sun's apogee, but less of it is stored.

Obviously, regions in similar zonal situations but lying at different latitudes can be very different according to the length of the day, phenological periodicity, maximum surface temperature, etc. These become controlling factors for the many specific landscapes in various regions of the tundra zone and play an important role in the distribution of plants and animals.

Relief and permafrost

There are different forms of relief: macro-relief (mountain ranges, lowlands and plains), meso-relief (hills, banks, terraces, sink-holes and depressions), micro-relief (hummocks, dunes, ravines) and nano-relief (small unevennesses in the ground surface such as tussocks, fissures and gullies).

The most important factor for the development of zonal landscapes is the macro-relief. A landscape may prevail in an alpine area of any zone analogous to that which will develop on a plain in a much colder climate. Vegetation similar to that of the tundra can, in principle, occur as a result of the effects of altitude within any zone on earth except for the polar desert.

In Eurasia the largest mountain massifs within the tundra zone are the Gory Byrranga and the Chukotskiy Khrebet. Within the limits of this zone there are in addition parts of the northern spurs of the Khibiny, the Urals and the central Siberian Plateau, Khrebet Verkhoyanskiy and the Khrebet Cherskogo as well as the Kolymskoye Nagor'ye. On mountain ranges situated within the tundra zone an analogue of the polar desert has developed: barren alpine rocky ground covered by a poor vegetation. On ranges lying beyond the limits of the subarctic, tundra extends far to the south.

From the Beloye More to the Olenek river the landscape is represented by plains with a quite broad structure. Its division into subzones is obvious and its borders more natural than those of other regions. The displacement of zonal borders towards the south or the north is here determined primarily by the increasing continentality.

East of the Olenek river numerous mountain ranges literally 'disrupt' the tundra zone and, locally, force it out completely. The effect of the mountain ranges east of the Kolyma river is especially strong. There the zonal tundra landscape is restricted to a narrow coastal belt only, forming a plain. On some maps the lowland parts of the coastal continent along the East Siberian and the Chukchi Seas are included in the arctic tundra subzone. Here the coastline runs north of the border of the typical subzone and the southern tundra is not represented at all within this region. The arctic subzone has developed on Ostrov Vrangelya as well as on the Novosibirskiye Ostrova.

Let us discuss the roles of 'meso-relief' and 'micro-relief'. When speaking of zonal characteristics within any habitat we have in mind primarily a relatively smooth, horizontal portion of a watershed landscape in which there is neither any underlying flow of ground water nor any significant erosion of the habitat. Such a locality is called a plakor [see translator's foreword]. The natural zonal type of vegetation is best developed on the plakor, for instance, a coniferous forest, a broad-leaved forest, steppe, and so on.

In areas of meso- and micro-relief where there is a subterranean flow of ground water, where the heating is more intense on south-facing slopes or where the drainage is bad, different kinds of intrazonal associations have developed: meadow, bogs, etc.

Two additional and important factors determine the specificity and the special variety of tundra meso- and micro-relief: cryogenic processes associated with the thawing of permanently frozen ground and ice cores, and a low amount of erosion due to the fact that the surface often consists of boulders, rocky substrates, stony cliffs, etc. The character of meso- and micro-relief varies strongly within the different subzones. In the southern tundra subzone and partly in the typical zone itself the relief is smoother and more gentle and the contours of the watersheds are represented by gently rolling outlines. In the northern subzone the outlines are broken, more uneven and with large scarps, especially rich in outcrops of rocks and boulders.

The following types of meso-relief are characteristic of the tundra zones: 1. watershed heights with smooth or slightly elevated ridges and very gentle slopes (this landscape element could be called a plakor but there is actually no typical plakor in the subarctic, analogous to the more extensive and flat territories at more southerly latitudes); 2. various kinds of boggy depressions between the watershed heights; 3. low, bog-like margins around large lakes, along rivers and by the sea (marshes); 4. river margins with steep and high banks or broad, sandy or gravelly spits; 5. ravines (ditches) and gullies caused by temporary water courses; 6. elevated river terraces and isolated remnants of banks; and 7. rocky outcrops.

The ridges and slopes of low hills are covered by vegetation characteristic of the subzone in question. More will be said about this later. In addition, the majority of the association types which can be called atypical of the tundra have developed here, such as different intra-zonal inclusions. The main representative is the mire [see translator's foreword]. It is very variable, sometimes even difficult to delimit from genuine tundra. Mires with polygons are to a great extent characteristic of the subarctic (see Fig. 9). They occur on low terraces, in lake basins and in drainage channels.

The surface of a mire can be broken by frost cracks into polygons (often tetragons), 7–5 m across. The central part of a polygon is flat or slightly concave and is bordered by an elevated rim, raised from 0.7 to 1 m above the surface of the centre. All the polygons are separated from each other by deep water-filled channels following along the cracks. A single mire can have from 50 to 100 such polygons. Mires with polygons are easily recognized from an aeroplane: their network of waterways stand out against the background of the slightly pitted structure of the flat tundra.

The hummocky mires are special. On a boggy surface, rows or groups of flat-topped peat hummocks, from 1 to 10 m in diameter, have developed, reaching heights from 0.5 to 1.5 m. They consist of peat formed by the mosses growing on top of them. The rows of hummocks are separated from each other by wet troughs and narrow water-filled spaces. This type of mire is more characteristic of the southern and the typical subzones in the western sector of subarctic Eurasia. Towards the north and especially in the arctic tundra they are less frequent. These flat-topped complexes must

Fig. 9. Diagram of the formation of polygonal bogs (a, viewed from above; b, horizontal profile). 1, Polygonal depressions, 2 and 3 raised borders and slopes, separating the trenches; 4, cracks separating adjacent polygons; 5, small pond in a depression; 6, upper water level; 7, upper level of permafrost.

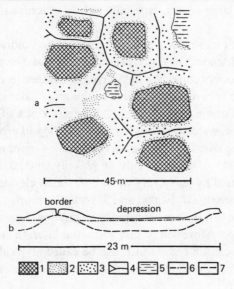

not be confused with the raised peat bogs of the forest tundra and the northern taiga, in which the height of the peat hillocks can reach 5–8 m and have a diameter of 25–100 m.

Another special type of orogenic micro-relief in the tundra is represented by the mound-like *baidzharakh* (see Fig. 10). *Baidzharakh* is a Yakutian term for relic mounds enclosing polygonal ice blocks. They are met with in groups or arranged in rows because of which they are also called 'rows of boxes' or 'burial mounds'. Their diameter is 3–10 m and they are 0.5–5 m tall, separated from each other by depressions. On gentle slopes in valleys of rivers and rivulets they can be seen as slightly projecting rows of mounds not more than 0.5 m in height. On steep slopes they often take on a conical shape and can become 1.5–3 m tall. Most peculiar are the circles of 'burial mounds' on steep slopes around hollows. On the bottom of such hollows runs a watercourse without any distinct bed: the water trickles between banks of grass and cotton-grass and, breaking through the walls of the hollows as a brooklet, it discharges into a larger brook. When neighbouring hollows are united, the 'burial mound' complex may occupy a very big area.

The geographical distribution of burial mounds is closely related to the development of polygonal ice inclusions which are found in north-eastern Asia and Alaska and in part where subterranean glaciation predominated over that above ground. The classical type of ice inclusions and thermokarst has developed primarily in the Novosibirskiye Ostrova.

A special element of relief in the arctic landscape is represented by the

Fig. 10. Baidzharakh hillocks on the island of Kotel'nyy (photo.: A. V. Krechmar).

'hydrolaccolithic mound'. In Yakutia these are called 'bulgunnyakhs' and in the Canadian Arctic 'pingos'. Unlike the burial mounds these are found in isolation. They can measure from 1 to 70 m in height with a corresponding diameter from 150 to 200 m. The pingos develop as a result of an underground ice lens deep in the ground. On its outside the pingo is covered with a layer of peat, about 1 m thick. Below it the frozen mineral ground consists of layered lake sediments, one to several metres thick. The mineral soil hides a cupola-shaped mass of ice. This is what determines the shape of the pingo.

It is not known what is below the ice of a pingo since at present no complete core has been drilled through the base of one. Local people have reported that sometimes during winter a 'deafening noise' ('explosion') can be heard near the pingo after which the ice heaves up above ground. Possibly there is a cavity below the ice, filled with water, perhaps in part with gas, which at a time when deep cracks develop in the surface of the pingo can escape to the outside. Well developed pingos have often a deep crack in the top, the formation of which starts the disintegration of the pingo. Along the fissure the vegetation cover dies off, then the peat and subsoil layers rip open, the ice is bared and starts to melt. From a distance the cracking mound appears double-topped. Pingos are related to boggy areas. Most often they are encountered in basins in habitats where lakes are being formed. As a rule the pingos occupy the lowest parts of the lake basins.

There are very conflicting opinions regarding the rate of pingo formation. Thus, V. N. Andreyev states that 'the process of pushing up a hydrolaccolithic pingo may last for hundreds, maybe a thousand years'. An American investigator stated when describing pingo mounds that they can appear suddenly during a winter, last for a few years and then disintegrate as the result of the ice layer melting. Under present climatic conditions the big pingos are found only in the taiga, the forest tundra and the southern parts of the tundra zone; in the typical tundra they are as a rule relic formations from the time when the climatic zones were displaced northward by about 2–3 degrees of latitude.

The most typical form of tundra micro-relief is, however, connected with cryogenic processes which cause either heaving or cracking of the ground. Thermokarst formation and erosion play the most important roles in shaping major forms of relief. The typical tundra landscape is a gently rolling plain, dissected by numerous watercourses and hollows, forming a network with a lot of small ponds, situated in groups inside large and low depressions.

The ponds appear as a result of locally thawing permafrost on level surfaces (see Fig. 11). Usually the slightly granular frozen sediment contains clean ice in the shape of a lens. When the ice has thawed, the ground subsides and forms a basin in which a pond may form. These ponds are usually not large, they may be up to 300 m long but only 3 m deep. The ponds are round, oval or roughly rectangular in shape; groups of ponds are often connected by meandering watercourses. In wide river valleys where a network of waterways has developed such lakes may be connected to a main river or, when drained, convert into mires.

Gullies are short but not very deep ditches with crumbling walls. They are found in valleys of large rivers but are not cut by tributary watercourses. During winter the gullies collect snow blown down from the surface of the watershed area; brooks forming in them during spring are fed by the melting snow but soon dry up. Under the influence of this water the gullies widen and form lateral spurs. Below the top surface the walls often crumble. An overhang forms over which water tumbles down in the spring. The tundra gullies are reminiscent of similar gullies found in the steppes.

In the gullies and along the banks of rivers and lakes (or more exactly, along their slopes) a vegetation has developed which is essentially different from that of the tundra. Many have written and talked about the colourful aspect of the tundra during summer, about the carpet of herbs and its multitude of flowers. Even in areas with a severe climate the monotonously species-poor tundra can display a riot of green hues and colourful flowers. The typical tundra plakor is usually uniform and poor in colours and has few bright and large flowers. But where there are ravines or steep slopes along the rivers a colourful and rich vegetation cover can be seen which

Fig. 11. Typical tundra pond with dense growth of cotton-grass.

sometimes reminds us of meadows in a forest or grassy and flowery roadsides: a dense sward of grasses with a multitude of flowering daisies, pinks, vetches, forget-me-nots, bistorts and many, many other less well known members of the temperate flora with their typically brightly coloured flowers. Above them colourful hoverflies stand still in the air and bumble-bees buzz. All this looks very similar to what we would call a meadow. It is, indeed, not a tundra but a meadow-like community.

Botanists are still arguing as to whether these associations of the tundra zone may be called meadows. They are often called a meadow-like tundra association. This would be the same as calling a glade in a forest a 'meadow-like forest'. It is, however, not a question about words but about the meaning of a natural phenomenon. The fact is that this type of vegetation does, in general, not occur in typical tundra. In these associations we find well formed humus soils with a close cover of grasses, the roots of which form a thick turf, and an abundance of herbs and few or almost no mosses. In addition, mixed variants are also found in which there are tundra characteristics (much moss, peaty soils, etc.). At the middle latitudes, trees often grow in meadows. Meadows are typical intrazonal associations. They can occur within any zone from the Arctic to the Tropics and almost always develop within the same kind of local biotopes, on slopes, flood plains or along lake shores, in the mountains and in habitats where the zonal vegetation cover has been obliterated, and so on. Their general characteristics are a predominance of mesophilic grasses (preferring an average amount of moisture) and dicotyledonous herbs.

There are habitats in our world where a meadow-like vegetation has spread from local biotopes into a plakor and occupies the major part of a watershed. Thus, in the European black earth (chernozem) forest steppe, two types of vegetation have developed on that kind of plakor habitat: predominantly aboreal or predominantly herbaceous associations which could be called meadow–steppes or steppe–meadows, respectively. They are reminiscent of true meadows with a closed grass sward and very many herbs. In contrast to the adjacent true steppe they do not dry out during the summer but continue to stay green and grow profusely until the autumn. Meadow-type vegetation forms where there is sufficient, but not too much, moisture. A balance between temperature and precipitation is characteristic of the climate on a steppe: the evapo-radiation index is almost balanced since about as much precipitation falls as can be evaporated by the heat due to the incident solar radiation.

In the tundra the evapo-radiation index is about 0.3. There is always a surplus of moisture there. However, in different local biotopes the conditions may be such that some enjoy more heat and less moisture. This can occur, e.g., on steep slopes with light soils. South-facing slopes within the tundra zone can heat up from one-and-a-half to twice the temperature of corresponding level habitats. But this does not necessarily mean that rich associations, typical of the tundra, will develop there. They are still subject to unfavourable and severe frosts during which they can be protected only by an adequately thick snow cover. Where snow is scarce, e.g., on top of slopes and at their edges, from where it is blown away by the wind, meadows will never develop. They will form only on slopes where much snow lies. However, too much snow, lasting too long, is not beneficial either.

Adequately thick snow protecting the soil from strong frost during the winter is a necessary condition for the formation of meadow-like vegetation within the tundra zone and because of this such 'meadows' are called nival or snow-bed meadows. Almost all these are developed along river banks below steep slopes. Animals, especially lemmings, play an important role in this formation by digging up and fertilizing the soil as do earthworms and other soil invertebrates which are more abundant here than in any other tundra habitat.

In the arctic tundra subzone the temperature is so much lower that even in the most favourable microclimates it is insufficient for a present-day formation of meadow-like vegetation. Here we encounter only fragments of herb–grass communities which – with some reservation – may be referred to as a type of meadow (see Figs. 12, 13). In the polar desert, meadow-like vegetation is never present.

The conditions for the development of meadow-like snow-bed vegetation are limited also in the southern part of the subarctic although the climate there is far more favourable. In the southern tundra subzone herb–grass communities can be seen everywhere but they are rarely as 'pure' as in the typical tundra subzone. They usually support many shrubs (willow or birch species), which form a most important element of the vegetation there. The shrubs in the southern tundra subzone have invaded the meadow-like vegetation from more favourable habitats.

The meadow-like tundra vegetation has attracted especial interest. As a result, it is possible to evaluate what the arctic tundra is capable of. These associations are richest and most productive under the prevailing conditions. The highest annual mass yield of the vegetation in the subarctic occurs there. In Taymyr a so-called meadow can produce on the surface up to 200 g/m² of dry matter per year, while the sedge–moss tundra yields no more than 25 g/m².

Besides the snow-bed meadows of the subarctic there are also other variants of meadow-like vegetation. For instance, the shoreline vegetation with hydrophilous grasses such as *Arctophila fulva* can also be included in this category but only in conditions such as maritime wetlands or along bays. Another variant of the meadow-like vegetation is formed as a result of activities of vertebrate animals, i.e., around fox dens and the 'feeding tables' of birds and also due to interference by man, e.g., around settlements, old camp sites, etc.

If on well drained slopes with sandy or gravelly soils there is only little snow accumulation, a meadow-like vegetation will not develop but instead a bush or shrub community and in habitats with very little snow cover a *Dryas* thicket. *Dryas* is one of the most characteristic of all arctic plant genera. It is a dwarf shrub with xeromorphic leaves, i.e., adapted to survive periods of insufficient moisture. The leaves are narrow with a smooth and leathery epidermis and involute margins, and the branches are spreading and woody. In the Eurasian tundra two species of *Dryas* are usually found: *D. punctata* and *D. octopetala*. In the northern belt of the subarctic, in the coldest and driest habitats, on edges of high hills and river banks as well as on gravel beds, the dryads form a dense and closed cover, tightly adpressed to the ground.

In the more southerly zone with such biotopes, very rich associations are formed consisting of berried shrubs of xeromorphic types such as cowberries (*Vaccinium vitis-idaea*), bog whortleberries (*Vaccinium uliginosum*), bearberries (*Arctostaphylos alpina* = *Arctous alpina*) and crowberries

Fig. 12. Meadow-like herbaceous vegetation on a south-facing slope of a ridge on arctic 'paved' tundra on Vaygach with the vegetation growing as long stripes.

(*Empetrum nigrum*). Characteristic of such habitats are the typical sub-arctic species reminiscent morphologically of succulents, such as the arctic bell-heather (*Cassiope tetragona*) and the native dwarf willow (*Salix nummularia*) with very small, leathery leaves. When somewhat more snow accumulates on the slopes, the shrub associations have an admixture of herbs and grasses, i.e., of meadow elements.

Some of the herbaceous plants are characteristic of dry, elevated habitats. They are xeromorphic in structure, having hard narrow leaves, sometimes very hairy. In such habitats grasses are found such as *Koeleria asiatica* and *Hierochloë alpina* and many others. Some species, typical of these habitats, form dense swards or tussocks. To these belong, e.g., the lousewort (*Pedicularis dasyantha*) and the oxytropes (*Oxytropismid-dendorffi* and *O. nigrescens*), the latter typical of skeletal soils.

When considering intrazonal inclusions within the basic zonal types of tundra landscapes, we can distinguish three main types: bog-like, meadow-like and shrub-like associations. Their geographical distribution is different. Bogs (mires) and meadows develop in principle within any zone on our earth. Consequently these associations are often called non-zonal or intrazonal. Similar examples are vegetations on sand spits and dunes, where for instance the beach grass *Leymus mollis* is characteristic. It is distributed along sea coasts from the Arctic to the Tropics. The xeromorphic shrubs and dwarf shrubs described above are met with mainly in subarctic

Fig. 13. Horizontal profile of vegetation on the arctic tundra in Taymyr. a, Meadow-like vegetation on a south-facing slope; b, grass–herb tundra.

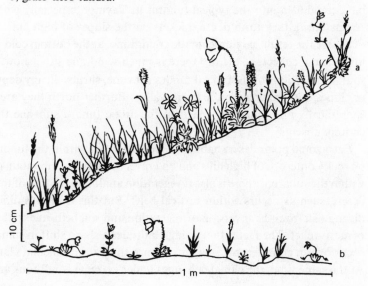

areas. Similar elements in a landscape are properly called intrazonal or 'intrastenozonal' (i.e., small associations within a zone; Chernov, 1975). Examples of such formations are saline and arid habitats.

Let us consider still another variant of intrazonal inclusions in the landscape. In the gullies of a steppe a meadow-like oak forest may develop or on high hills in the north an open woodland of boreal type in tundra habitats. Under the prevailing conditions, the mesophilic oak forest and the tundra woodland are found outside 'their proper zone' (i.e., broad-leaved forest and coniferous forest, respectively). Such landscape elements are called extrazonal. They are usually interrelated with slopes, depressions or ridge tops. The mechanisms forming extrazonal vegetation are included under the so-called 'law of preference', which means that plants, usually inhabiting a plakor and forming zonal associations, have spread in the south onto north-facing slopes and depressions, where it is cooler, and in the north onto warmer south-facing slopes. Within the limits of the tundra zone there may be found extrazonal habitats supporting a vegetation of southern zonal type such as forest-like associations, or of a northern type such as polar desert associations.

In the southern portions of the tundra zone everywhere there are forest islands situated on river terraces and on south facing slopes of the watersheds. In Taymyr and in the Olenek river basin these reach as far north as latitude 72 °N. In the typical tundra subzone, we often meet with thickets of willow, birch and alder. These are always extrazonal elements. Usually such associations form on hillsides and lake beds and where snow accumulates. Shrub-alders (*Alnus crispa*) are typical of the southern tundra but extend also into the typical tundra. In Taymyr, especially within the Agapa basin, they form dense thickets on the slopes of high hills. This is of course the result of microclimatic conditions: at the bottom cold air and much snow collect, at the top there is much wind and little snow. In the southern portions of the typical tundra subzone, shrubs fill any depression or slopes and are common in every valley. Further north they are found sporadically and more dispersed; in the arctic tundra subzone they are entirely absent.

Extrazonal polar desert habitats have developed within the tundra zone on rocky outcrops of high hills and on gravel beds. In the mountain massifs within the subarctic there is also a vegetation analogous to that of the polar desert, such as occurs within vertical belts. But this is always difficult to distinguish from the inversion of vegetation and soil belts due to the effect of meso-relief. The fact is that at high latitudes a sharp shift in climate can occur over a considerably smaller distance than at intermediate latitudes. At the same time, the climate changes under extreme conditions and may affect soils and vegetation to a lesser extent. Because of this, elevation or low relief is not important. Neither is the shift from south to north strongly indicative of the aspect of the landscape.

Within the tundra zone there are very many different forms of *nano-relief* as a result of the cryogenic processes that occur in the ground layer when thawing in summer. The thickness of this layer varies from 30 to 200 cm and depends on exposure and gradient, moisture conditions and mechanical composition of the soil, thickness and duration of the snow cover as well as on the character of the vegetation. In the southern subarctic, the depth of the seasonal thaw in a plakor habitat is 80 to 90 cm, in the typical subzone and the arctic subzone 40 to 60 cm, and in the polar desert 20 to 40 cm.

Depending on the weather, the difference in depth thawed during different years may amount from 10 to 20 cm, because of which it is possible to distinguish three layers: the top layer thawing out every year, the periodically active layer thawing only during years of higher temperatures, and the layer permanently frozen under present-day conditions. Because of this, roots of plants can penetrate into otherwise frozen soils during summer and in mild years. Mineral and well-drained soils thaw more quickly and to a greater depth than peaty and wet soils. The thawing continues throughout the summer but even as late as the end of July a frozen surface can be found at a depth of only 20–30 cm. The thickness of the active layer depends on the exposure and the gradient of the surface. Within the typical tundra subzone, e.g., in Taymyr, the depth of the seasonally thawed ground on the plakor is about 50 cm, but on south-facing slopes it is 150 cm. Further north the difference in warming between flat surfaces and south-facing slopes is reduced. In the arctic tundra the active layer both on watersheds and other slopes may be only 40–60 cm thick.

During spring, summer and autumn, miscellaneous cryogenic processes occur in the active layer: heaving, crack-formation and sorting of the ground, stones being squeezed out of fine soil, solifluction (slow flow of soil masses saturated with water), expulsion (watery soils spilling out over the surface) and other types of action. All this leads to the formation of various forms of tundra nano-relief which in the specialized literature is called 'patterned ground'. This collective term mainly covers the symmetrical forms of the arctic nano-relief: circles and polygons, medallions, networks and stripes, etc.

The American scientist, A. L. Washburn, in his basic classification of nano-relief (Washburn, 1958) distinguished elemental forms, and the presence or absence of sorted ground. For instance, circles are structured ground in the form of variously developed circular cells. These are distinguished into sorted or non-sorted types depending on whether they are bordered or not by stone blocks (see Fig. 14). Stone circles and stone rings can be classified as sorted circles. They are found both individually and in groups and have a diameter varying from 0.8 to 30 m or more. The centre portion of the majority of sorted circles is usually covered with fine soil but sometimes with coarser gravel. Sorted circles are characteristic of the polar desert and rarely met with in the subarctic or alpine areas. Usually they have no vegetation although sometimes isolated specimens of plants grow within them.

By non-sorted circles the American author means formations without a block border. These may also occur in groups. In the Russian literature such habitats have been given various terms such as 'spotted' or 'medallion'

Fig. 14. Distribution of blocks and flat stones within a 'sorted circle'.

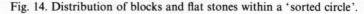

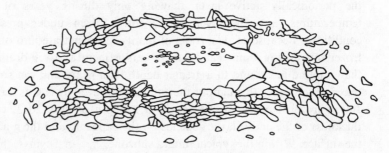

Fig. 15. Profile of a patch of bare ground in the typical tundra subzone. 1, Moss sward; 2, moss peat; 3, surface of bare ground; 4, surface of permafrost.

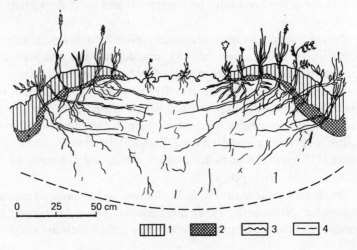

tundra (see Figs. 6, 7, 15), i.e., patches of bare ground with sparse vegetation, sometimes covered with clumps of crustaceous lichens and surrounded by raised borders with a well-developed moss carpet, some 5–10 cm above ground level. Between the borders of two adjacent spots a depression occurs in the form of a shallow trench filled with peat and mosses mixed with flowering plants.

The polygons are patterned ground in which the cells actually take the shape of straight sided polygons. These are never found singly. They may measure from a few centimetres to tens of metres across. The sorted polygons have borders of big blocks around a fine soil. The Soviet tundra specialists call such structures stony polygons, stone nets or polygonal soils. The sorted polygons are mainly distributed in the polar desert and in mountain tundras. The non-sorted polygons do not have block borders. There are also polygonal tundras and polygonal mires of which more will be said later (see Fig. 9).

The stone nets are a form of patterned ground composed of circles or polygons. They, too, may be sorted or non-sorted, i.e., with or without block borders. The vegetation occurs in the cracks between the cells and forms a network. The basic type of non-sorted net is plakor with low-hummocky tundra in which the nano-relief is represented by small hummocks, 10–30 cm tall, separated by shallow trenches. Usually a closed moss carpet has developed on this tundra. Such a type of patterned ground is characteristic of the southern and typical tundra subzones and hardly ever met with in the arctic tundra proper.

The spotted and polygonal complexes are astonishing, not only to new-comers but also to those who have visited the Arctic many times. They present a really surprising picture where, set in the solid cover evenly spaced patches of bare ground appear looking like 'windows' or 'eyes' in the tundra; it is not without reason that in the Canadian literature these bare patches have been called 'orifices of the earth'. The dimensions of the patches, their shape and number in individual areas depend on the mechanical composition of the ground (in sandy soil they are smaller, in loamy soil larger), on the relief of the ground (on a sloping surface they have an ovoid shape, on a horizontal area a more circular form). Towards the north they are smaller, but more frequent (see Fig. 16). In Taymyr there are about 15 bare patches in an area 10×10 m, but in the typical tundra on a similar surface there may be up to 30 and in the arctic tundra more than 100. Not only do the dimensions of the patches diminish but also the width of the moss borders between them: from 1 to 2 m in the southern tundra to 20 to 30 cm in the northern.

The causes of the formation of these patches of bare ground are manifold. Usually they appear in the southern tundra areas as the results of a break in closed vegetation.The patches which appear are gradually refilled with vegetation as a result of which closed cover is restored. It is true that this process is very slow. It may require decades or even hundreds of years. In the northern tundra zones and in the polar desert the process forming bare patches is different. The plants colonize cracks in the frozen ground where snow collects during winter. As a result a vegetation cover develops in the form of a network interspersed with cells of bare ground (see Figs. 17, 18). Under the conditions of the present severe climate there, the vegetation achieves this appearance extremely slowly.

Even during the early stages of investigations within the tundra zones the problem regarding the origin of patterned ground aroused the interest of many scientists. The Russian botanists V. N. Sukachev and B. N. Gorodkov paid great attention to this problem. A large number of hypotheses have been advanced regarding the formation of the spotted and polygonal types of tundra. While in the beginning these ideas were aimed at explaining the operation of every single factor, at present almost all lean toward polygenetic concepts and, thus, the origin of patterned ground is considered to be associated with various aspects of cryogenesis.

A number of hypotheses are based on the appearance of cracks in the ground, resulting from frost action. Because of this, for instance, the sorting of blocks from finer soils can be explained. When the soil freezes, maximum contraction of its mass occurs in habitats with the highest concentration of fine soils. Under extreme conditions this surface cracking may even penetrate to the rock beneath. At the time of thawing and compression, the fine soil contracts as a result of cohesion, leaving the blocks where they were found originally. Repeated over and over again this process results in the blocks, so to speak, being pushed up onto the surface of the ground (see Fig. 19) and the fine soil filling in under it. This is the explanation of sorted material in patterned ground.

Fig. 16. Diagram of the distribution of patches of bare ground in: a, typical tundra subzone and b, arctic tundra subzone, (5 m × 5 m square).

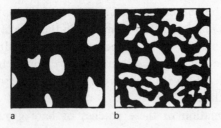

a   b

Fig. 17. Diagram of the formation of polygonal, spotted arctic tundra. a, Frostcracks in bare ground; b, moss cover over adjacent edges of the polygon; c, old spots: in place of the cracks wide gulleys filled with accumulated peat and a moss sward. 1, Mineral soil; 2, moss sward; 3, peat; 4, 'spot' or 'medallion'; 5, frost crack.

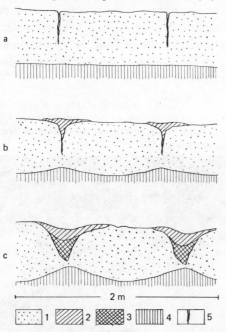

Fig. 18. Three variants of polygonal ground with gradual development of the moss sward over the frost cracks. a, Flat patches with minimal vegetation cover; b, net-like formation of vegetation growing in widening cracks; c, still denser vegetation, filling out the progressing and widening cracks.

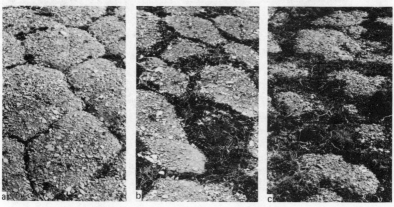

Ideas regarding the mechanisms forming certain varieties of spotted (or medallion) tundra are based on cracking and heaving of the ground. In habitats where the vegetation cover is thin or where there is little snow, the ground freezes quickly and may locally heave because of expansion. As a result the surface curves upward. If the process exerts a persistent pressure, the sward may rupture. During subsequent snow, water or wind erosion, the plants are completely destroyed and, as a result, bare ground appears. The spotted tundra in the southern parts of the tundra zone owes its formation to just this process.

Fig. 19. Cryogenic phenomena in spotted tundra. a, Frost-cracks, flat stones and small roots on the surface of bare ground in the typical tundra subzone; b, willow roots, resting on the surface of bare ground in the southern tundra subzone.

As a result of the ground freezing from the surface toward the top of the permafrost layer, a high hydrostatic pressure develops in the still unfrozen but very wet and muddy soil. As a result it may ooze out over the surface through some cracks. V. N. Sukachev in 1911 based his hypothesis regarding the formation of patches of bare ground on this kind of assumption. It has been discussed repeatedly whether, in the majority of the cases, this is a catastrophic event in the form of a sudden eruption of fluid soil. But really it has not been fully explained how the half-fluid material can suddenly be ejected on to the surface. However, the hypothesis agrees well with the phenomenon of so-called 'thixotropic' tundra soils. Thixotropism is the ability of a relatively thin layer of soil to convert into a thin, fluid state under the influence of additional load. Those working on the tundra know that, by stamping rhythmically on the ground, it is possible to make an area of several square feet of the ground quiver. When the pressure is relieved, the ground becomes solid again. A similar idea may be suggested for the cryostatic pressure as well. No doubt, some forms of spotted tundra are formed in this manner (Govorukhin, 1960).

Some hypotheses regarding the formation of patterned ground are based on the appearance of pressure as a result of drying out and the effect of low temperatures. During summer when the ground is drying out, cracks will develop in it which fill with water and widen later when the water freezes. Where found, such a system of cracks will usually spread further. This could explain the development of minor polygonal structures.

According to another hypothesis the polygons form due to stress within the frozen soil at very low temperatures. B. N. Gorodkov considered frost cracking as the principal cause for the formation of polygonal arctic tundra. If there is much ice in the ground a frost crack may form as result of stress due to temperatures well below freezing which lead to the formation of an entire polygonal system. This could be the explanation of the origin of polygonal mires, with polygons up to 10 m in diameter.

Thus, the genesis of a particular type of patterned ground depends on different processes. Polygonal structures of minor dimensions develop at high latitudes due to the effect of stress when drying out and because of low temperatures. During the formation of circular forms, most common in the middle belt of the tundra zone, differential heaving assumes great importance. Sorted ground occurs due to the ejection of blocks following alternating freezing and thawing. It is possible that the processes of pressure, uplift and expulsion from the ground may be combined. In order to explain fully the processes of the formation of present as well as of fossil patterned ground more detailed and exact information is necessary. Experiments in cold chambers and field-work at different times of the year are much needed. Co-operation is necessary between research workers from the fields of geology, physics, cryogenics, soil science and botany.

The appearance of the tundra landscape is more or less associated with permafrost and the processes governing its seasonal and perennial dynamics. Its effect on the vegetation may, on the whole, be judged as negative: the drop in soil temperature, the soaking of the ground, the mechanical impediment to development of the root system, the damage done to the vegetation cover due to the different kinds of cryogenic processes, etc. All this disturbs the normal activities of the plants and limits their access to necessary materials. The less the depth of the seasonal thaw, the poorer is the vegetation and the larger the area of open ground (see Fig. 16).

However in its turn the vegetation cover also exerts an effect on the depth of the thawed ground. Between the permafrost layer and the vegetation of the tundra there is always a complicated and contradictory inter-relationship. The thicker the vegetation cover, especially when it is formed by a closed moss carpet, the greater is its heat insulating properties and the lower the soil temperature. Consequently the top of the permafrost layer will be found nearer the ground surface and thus the conditions for the development of flowering plants and their root systems are worse. Denuded soils, on the other hand, heat up better. This is easily observed in the spotted tundra. During spring the upper limit of the permafrost quickly becomes lower under the bare patches but more slowly under the thick moss carpet in the surrounding cracks. It is only at the end of the summer that the upper surface of the permafrost becomes level under all the elements of the spotted tundra. Well-heated bare ground is actively invaded by various plant species, but with the formation of a closed sward the heating of the soil gradually diminishes, again allowing the permafrost surface to rise. One of the causes for the colourful aspect of the arctic tundra, the abundance of flowering plants, depends on the fact that under these climatic conditions a closed and thick moss carpet cannot form and therefore the soils heat up quickly and are well aerated.

When the vegetation cover is disrupted, thermokarst often starts to form, i.e., the ground ice melts and the surface subsides. Depending on the amount of ice and its level in the ground the thermokarst can assume different dimensions. If it occurs over a small area, deep water-filled troughs develop, becoming colonized by hygrophilous (water-loving) mosses, especially by *Calliergon* species, gradually filling them and converting them into small saucer-shaped mires. The growth of the hygrophilous mosses forms a thick cushion, which better insulates the ground against heat reaching it and the level of the permafrost rises again. Instead of the small mire a typical tundra vegetation may develop anew.

Thermokarst is often caused by construction work disrupting the vegetation cover. Destruction of the vegetation cover is caused especially by tracked vehicles, so widely used for transport in the northlands. The tracks of the cross-country vehicles tear up the sward, exposing the mineral soil and starting thermokarst formation as a result of which the land subsides and gullies form.

The vegetation cover and the permafrost are components belonging to a single system which in every stable habitat can be found in a state of some kind of equilibrium under the particular conditions of the climatic regime. The stability of this equilibrium is a token of the orderly structure of the tundra landscape.

5

Snow and its role in the life of the tundra

Snow cover is a very complex physico–geographical phenomenon. It is characterized by its thickness, density, heat conducting capacity, albedo, duration and the structure of the particles composing it, etc. Its thickness (or quantity) is most important as well as the duration of its cover. There is, however, no direct relationship between these qualities. Sometimes a covering of snow persists longest where it is quite thin. In the tundra zone the snow lasts from 200 to 280 days, much longer than within other zones. In the middle belt of the European parts of the USSR it lasts for 120–160 days, on the steppes 60–80 days.

The thickness of the snow cover within the tundra zone varies between 10 and 50 cm but in most places it is 20–30 cm thick, about the same as in the forest tundra. The thickness of the snow cover in no way diminishes relative to latitude or zonal borders. In Siberia, including also the subarctic territory, there is a considerably greater change from west to east than from north to south. This is related to the gradually increasing continentality and the normal amount of precipitation, as well as to its distribution during the course of the year. In the Atlantic parts of the subarctic, on Poluostrov Kanina and on the Malozemel'skaya and Bol'shezemel'skaya Tundras, the snow cover is about 50 cm thick but in the east from the northern parts of Yamal to Chukotka rarely more than 20–30 cm. The duration of the snow cover is, on the other hand, maximum in the central and eastern parts of the tundra zone and reaches a minimum closer to the Atlantic. On Taymyr the snow stays from 260 days in the south to 280 days in the north, in the Bol'shezemel'skaya Tundra and on southern Novaya Zemlya about 240 days; in the northern part of the Kol'skiy Poluostrov it lasts about 200 days, that is, almost as long as on the steppes of eastern Siberia. However, it is sometimes difficult to distinguish what is characteristic of a zone from the effects of continentality or proximity to an ocean.

In the western parts of the subarctic close to the Atlantic and the modifying effects of the Gulf Stream, the duration of the snowy season is especially short. As a consequence of frequent precipitation there during winter snow-beds may form on hill slopes, among rocks and in narrow valleys. These beds may not melt the same year. When accumulating over a period of some years, the snow compacts, changes into firn and later on into ice. This process occurs in areas of present day glaciation, e.g., Greenland.

The following qualities are typical of the snow cover on the Siberian tundra: in the southern subzone it is about 50 cm thick, in the typical tundra *c.* 25 cm and in the arctic *c.* 10–20 cm. Due to the characteristic thinness of the layer in the tundra, every single rise in the micro-relief – even the most insignificant – is bared during winter and early spring and subjected to the prolonged effect of low temperatures. However, on the tundra the snow, although light, lasts for a long time. This is another negative factor in the development of vegetation and in the living conditions of the animals.

A very important quality of the snow cover in the Arctic is its uneven distribution. The snow on the tundra is very 'dry' and powdery. Constant strong winds blow it away from the ridge tops and banks and a layer, only 2–5 cm thick, may be left there which does not even cover the low-growing plants. However, in depressions, in gullies and on slopes the snow may form thick drifts. Due to the wind effect the snow becomes strongly compacted. By December it can support a man without skis. The longer it lasts, the harder it becomes, compacting so hard that a snow-plough can barely penetrate it. Sometimes especially dense surface layers are formed (snow crusts) alternating with softer layers. Even during spring, the period of most intense thawing, it may be possible to walk on the snow drifts without sinking in.

The great density and the excellent insulating characteristics of the snow in the Arctic makes it useful for building igloos. These have been best developed in the American north and on Greenland. In an Eskimo igloo, constructed from snow blocks, it is possible to maintain positive temperatures by means of a small stove even at external temperatures around $-40\ ^\circ$C. The Eskimos manage very cleverly in not allowing the snow to melt. Because of the great reserve of cold in the thick snow walls, the inside walls of the igloo melt only slowly when heated. The water drops freeze, covering the inner wall with an icy crust which improves the durability of the snow hut.

The role of the snow takes many forms on the tundra: it participates in the development of a heat regime, especially in terms of reflection of solar radiation (albedo) and absorption of heat when thawing; it reduces the processes of erosion and denudation; it protects plants and animals from the cold during winter; its grains can be corrosive; it limits the period of active life, etc. As a heat insulator snow is of great importance in protecting the soils, the vegetation and the animals from the low winter temperatures. The heat insulating role of the snow depends on its capacity for heat conduction but ultimately on its density, i.e., on the amount of air contained in it. There is a minimum of heat conduction in fresh snow, falling as big flakes during calm and relatively warm weather; the maximum is found in strongly compacted snow and firn. The difference in density between freshly fallen and strongly compacted snow is tenfold, the heat conduction a hundredfold. Melting snow is two to three times more heat conducting than dry snow of the same density (Rikhter, 1945).

Heat exchange between the snow covering the ground and the atmosphere is almost nil. Daily fluctuations in temperature are found only down to a depth of 25 cm into the snow. Short term changes in air temperature are clearly reflected in a layer 0–5 cm deep only. At middle latitudes the ground may remain free from frost throughout the winter under a 50–70 cm thickness of snow. Positive temperatures may be found even under a 20 cm thick layer. However, during winter on the tundra, of course no soils remain free from frost. Although the snow quickly covers thawed ground, this will freeze as the cold also penetrates from below. In permafrost areas the temperature at a depth of 1.5–2 m is about 1–3 °C below freezing. Throughout the winter, conditions below the snow are quite tolerable not only preserving animals and plants during the resting stage, but even allowing active life of warm-blooded animals such as lemmings, voles, stoats and weasels. Actually, the lemmings and the tundra voles breed successfully at this time. T. N. Dunayeva (1948) has studied the life of tundra rodents during the winter.

Middendorff's vole (*Microtus middendorffii*) and the narrow-skulled vole (*M. gregalis*) are distributed mainly in the southern tundra subzone where the snow cover is always thick but light. The Siberian lemming (*Lemmus sibiricus*) and the collared (hoofed) lemming (*Dicrostonyx torquatus*) inhabit all the subzones of the tundra as well as the polar desert. In areas where the snow is scarce, they move during winter to habitats in thicker snow drifts, on slopes, in river valleys and in depressions between banks of boulders. The collared lemming is, indeed, better adapted to a life in compact snow. It has developed peculiar forked claws (hoofs) which, it is said, facilitate the digging of tunnels. During calm weather it is possible to hear the grating sound as this lemming scrapes through the snow.

During the course of a long winter the lemmings dig durable tunnels through the moss carpet to a radius of 10 m or more from the nest. They feed mainly on fleshy underground parts of sedges (*Carex*) and cotton-grass (*Eriophorum*) growing in the thick moss sward. In order to reach them, the little animals 'cut away' enormous quantities of moss and excavate and gnaw out long tunnels in the hummocks. They do not consume the dry parts of the sedges and the cotton-grass but use them for constructing their ball-shaped nests deep within the snow. Based on the distribution of the nests when melted out, it is possible to learn much about the life of the lemmings during the winter. The nests are located at a fair distance from each other. Even during years when the lemming density is high, the nests are about 50 m apart, locally 100–150 m. Indeed, this is determined by the quantity of roots available for fodder. Some two to three nests may be found at only 2–5 m distance from each other; these, of course, belong to the same family. The nests are located only where a long lasting and adequately thick snow cover – not less than 50 cm – develops early; in depressions between hills, on slopes of valleys and at the foot of cliffs, in mires between peaty hummocks, and along river banks, etc.

The time of the most frequent reproduction of the lemmings under the snow is not known exactly. It is believed to occur, in general, from April to June. There are data on the birth of young as early as January or February. An indication of this is furnished in particular by the discovery of young animals in the stomachs of foxes, caught during the winter. Mid-winter propagation is more characteristic of the Siberian lemming and is actually related to the better thermoregulation of that species.

Stoats (*Mustela erminea*) and other weasels also live under the snow on the tundra. They hunt voles and lemmings by taking advantage of their burrows, tunnels and nests in which abundant remnants may be found of meals enjoyed by these predatory animals. They avoid the watersheds on the tundra and prefer to live in ravines and on hill sides where there is a greater quantity of less compacted snow.

Snow is also the most important factor in the winter life of the large herbivorous mammals and birds such as the reindeer (*Rangifer tarandus*), the musk ox (*Ovibus moschatus*) and the snow hare (*Lepus timidus*) as well as the willow grouse and ptarmigan (*Lagopus l. lagopus* and *L. mutus*), (see Figs. 20, 21). Somehow they must all obtain their food from the vegetation hidden under the snow. In the southern half of the tundra zone the hare consumes branches of bushes jutting up above the snow during winter. There are few hares in the tundra and this scarce and coarse food seems fully adequate for them. But for the reindeer, the ptarmigan and the grouse the food is insufficient. They are not able to tear up a thick layer of very compact snow, and therefore in the autumn migrate southwards into the forest tundra and the taiga, where the snow is loose and food more plentiful.

However, locally, reindeer exist the year round in the most northerly parts of the tundra zone, e.g. on Belyy Ostrov, Novaya Zemlya, Novosibirskiye Ostrova and Ostrov Vrangelya. The conditions at the very high latitudes are of course very severe and food is very scarce. However, because there is only scanty snow cover in the arctic tundra, there are many patches blown free of snow where the reindeer can find food even during the winter. At that time of the year their food consists of willow twigs, juicy

Fig. 20. The ptarmigan (*Lagopus mutus*), a typical inhabitant of tundra and mountain nival landscapes (photo.: A. V. Krechmar).

Fig. 21. Behaviour of ptarmigan when digging into snow (according to Andreyev, 1975). 1, Selection of locality; 2, the digging starts; 3, a tunnel is dug; 4, the bird surveys area, lifting its head above the surface of the snow; 5, position of the ptarmigan under the snow when it is being hunted.

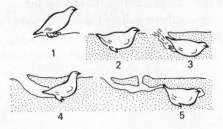

roots and runners as well as overwintering buds of cotton-grass (*Eriophorum*), saxifrages, grasses, etc. (Kishchinskiy, 1971). It may seem surprising that they can survive on such poor and scarce vegetation during the severe winter. It is, however, a fact that in the arctic tundra the reindeer are evidently able to survive more successfully than in the typical tundra or the southern tundra subzones.

The willow grouse (*Lagopus l. lagopus*) generally inhabits the southern half of the subarctic areas. During winter it migrates to the forest tundra or the taiga. There are, however, indications that sometimes it stays in the arctic tundra, e.g., on Novosibirskiye Ostrova, although it does not usually nest in that subzone.

Only one species of large herbivorous mammals has succeeded in adapting really well to the living conditions in the arctic tundra and even to those in the polar desert. This is the musk ox (*Ovibos moschatus*) which has never been widely distributed anywhere in the Arctic and now is met with in the wild state only in the American sector. Thanks to the scanty snow cover in the arctic tundra, this large mammal is able to survive there permanently.

What role the snow plays in plant life can be seen from, e.g., the distribution of bushes on the slopes of a ravine. The tallest bushes grow on the middle or at the foot of the slope. At the top they become adpressed to the ground and literally 'trimmed off'. During winter or early spring when the thickness of the snow is measured, it is evident that it reaches exactly the same height as that of the bushes. The cause is not only the warming effect of the snow cover. Often bushes of birch (*Betula*), willow (*Salix*) and alder (*Alnus*) are encountered with dry tops from which the bark has been stripped due to snow abrasion. These branches had dried out and gradually died off. The surface of the snow determines the size of the bushes since their top buds are damaged. This in turn leads to increased branching of the shoots. That is why bushes in the tundra have the appearance of having been constantly trimmed.

In the southern tundra, larches (*Larix*) may be met with having the appearance of strongly branched bushes pressed flat to the ground and reaching a height of only 30–50 cm. This growth form is called 'krummholz' (from the German word for crooked wood), a shape which is assumed by many tree species in the subarctic. Sometimes they form thick, almost impenetrable thickets. Krummholz is especially characteristic of mountain areas and the northlands of the Far East where the tundra landscape extends to very low latitudes and into areas of many tree species. There, species of the pine family (*Pinaceae*) are widely distributed as krummholz. This is valid for other tree species as well. In the krummholz thickets

conditions during winter are very favourable for animals: under the thick
snow cover lying on top of the branches there is a multitude of empty
spaces, i.e., habitats within a layer of litter and soil. These sites allow both
movement and access to food.

The height of the bushes is not always determined by the thickness of
the snow cover. The alders (*Alnus*) in the southern tundra may, for
instance, reach a height of 3 m with their tops just up above the snow
during winter. Their profile is, however, remarkable because of the
arrangement of the branches. Above the dense many-branched bush,
hibernating below snow, straight stems push up deprived of their side
branches. At the very top they branch out again (see Fig. 22). Similar
'double-crowned' shapes are often met with as well in larches. They are
called 'trees in skirts'. The crown shape is a result of the fact that the snow
abrasion is especially strong just at the surface of the snow layer. A little
higher up, where it is not as intense, slender branches can grow out and
form a secondary crown, sometimes called a 'flag'. Larch, spruce and birch
often display so-called 'stilted' growth forms: low-growing trees standing
on a number of distorted and denuded thin stems.

The snow cover prevents processes of erosion and denudation during
winter as well as the transport of mineral particles. But in the tundra and
especially in the northern subzone there are many places where the snow
is completely absent and the soils are bare, especially on ridge tops and
along the edges of river banks. These parts are subject to an intense snow
abrasion. Bushes growing there have strongly denuded, adpressed shoots
on which the bark has often been polished off. If the soils on such heights
are light, sandy or loam–sandy, they will locally be blown or transported
away, exposing the roots of the plants which then die off. In such places

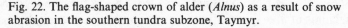

Fig. 22. The flag-shaped crown of alder (*Alnus*) as a result of snow
abrasion in the southern tundra subzone, Taymyr.

so-called 'blow-outs', hollows reaching a depth of 1 m, are found. Habitats with clay, loam or stony soils are more stable. Often an unbroken cover of evergreen *Dryas* forms on them, resisting winter denudation thanks to their resilient and dense leaves.

Habitats with thin snow cover disappearing early in the spring are subject to a special effect of the severe arctic conditions, so-called 'frost-boils' (a term used by American authors). During the first sunny days of spring such places thaw out temporarily but freeze again during the night when the temperature drops. They crack and in the fissures and the openings in the soil ice wedges, crystals of ice or 'ice fingers' form. Simultaneously vertical movements occur in the soil mass. Soil that is alternately freezing and thawing will lift and heave. As a result of these processes, the surface of the soil acquires a lamellated, or porous structure having the appearance of a boiling liquid.

The main reason for the extreme animal–vegetation relationships within the tundra zones is the short growing season. It is determined not only by the radiation conditions and the shortage of solar heat but also by the duration of the snow cover, which absorbs and reflects the solar radiation. The snow will start to melt on the tundra although the air temperature is still negative. Locally it will have already disappeared from ridge tops and the edges of river banks by the middle of May. On the southern tundra the snow starts to melt faster at the end of May and on the typical tundra at the beginning or the middle of June. The speed of its disappearance depends on the micro- or nano-relief (see Fig. 23).

Fig. 23. Spring in the polygonal tundra. Snow lingering at the level of the raised borders of the 'medallions' and in the depressions between.

At the end of May or beginning of June there is usually some clear weather with low humidity. This may be followed by strong winds, sometimes a blizzard, which in turn are followed by a warm period and a very intense thaw. Often the snow disappears within three to four days. Thus, for instance, on 5 June 1967 in Taymyr, about 95% of the tundra was covered with snow. On the 7th a thaw occurred and by the end of the day only 50% of the snow remained. By the 8 June just 10% of the tundra was covered and on 9 June only 2% remained covered.

However, the snow does not disappear that fast every year and not everywhere on the tundra. Often the thaw comes late in the spring. In depressions, on slopes and in stream valleys the snow can remain into July and when it is very thick, even into August.

The inhabitants of the tundra try to make as effective use as possible of the period suitable for active life early in the spring when the solar radiation is high, but the air temperature is still low and much of the snow remains. Life on the tundra during the spring develops very vigorously and depends on the rate at which the snow melts, the distribution of thawed and snow-covered patches, the meltwater runnels, the frequent snowfalls and the different physical characteristics of the snow. In general at this time the snow plays a major role in the life of all groups of tundra organisms. As the season proceeds it becomes especially important both for the vegetation and the animal populations at what speed and in what sequence the snow disappears from the different parts of the surface.

In descriptions of the vernal period of tundra life one often comes across mention of the so-called 'snow hot-houses' which play a great role in the development of the vegetation (Tikhomirov, 1963). 'Snow hot-houses' are small cavities between the soil and the lower surface of the snow, formed as a result of heating of the soil by the solar radiation penetrating through the layer of snow. The lower layers of the snow then melt leaving a small dome. The snow above this cavity crystallizes and turns into ice covering this miniature greenhouse like a window. Inside there is a thermal effect – as it is known in physics – in operation: entering the cavity through the transparent cover the solar rays heat the deep-lying darker particles. The thermal re-radiation caused by this is repeatedly reflected by the 'window', raising the temperature inside by as much as 10 °C above that of the outside air. Usually the heat surplus amounts to 2–5 °C. The 'hothouses' take many shapes; some of them are covered by a thin transparent sheet of fused snow crystals, others by thick, semi-transparent layers of frozen snow. Some form around the plants shoots or leaves which will then penetrate the cover.

These 'hothouses' form in early spring when the air temperature is still below 0 °C. They are typical of the arctic tundra subzone where the snow cover is thin and, when melting, gives rise to a multitude of snow-free patches alternating with small snowdrifts. Indeed, in this subzone they are of decisive importance for the development of the plants by allowing them to start growth even during a period when air temperatures are below freezing. Aleksandrova (1961) observed the vegetation on Bol'shoy Lyak-hovskiy Ostrov and found inside such 'hot-houses' specimens of chickweed (*Cerastium*), cinquefoil (*Potentilla*), saxifrages, and poppies (*Papaver*) during the early part of June growing while the air temperature was below 0 °C. Inside the 'hot-houses' the plants, green during winter, develop an anthocyanine colour, start to produce new shoots and even to bud.

Thus, thanks to the thermal effect, the plants are indeed able to utilize the solar radiation for their vegetative stage even at air temperatures below zero. During the period of maximum snow melting and when the soils are being heated further, the 'hot-houses' gradually lose their importance. The plants begin to grow vegetatively at an accelerated rate on snow-free patches, but where these previously had 'hot-houses' the vegetation is able to develop at a much faster rate.

South of the Arctic Tundra Subzone where there is a much thicker snow cover, melting is of the evaporation type, i.e., ablation with water vapour condensing back onto the surface and bare spots appearing simultaneously over large areas. Under such conditions 'hot-houses' do not of course form and are of no real importance for the plants. However, in general, the role of the thermal effect for the seasonal development of tundra vegetation is not entirely clear. The problem needs more detailed research and specifically designed experiments. The frequently discussed lesser or greater roles of the 'hot-houses' for the life of the tundra plants has been confirmed only on the basis of short-term observations by naturalists. It is believed that the attractive aspect of this particular phenomenon affects the positive appraisal of its importance.

The thermal rays that penetrate beneath the snow are definitely of importance for the life of the tundra macrophytes, i.e., the shrubs. In the typical tundra subzone, willow (*Salix*) shrubs grow on slopes and in valleys where large snow-beds form, which do not melt until late in the season. The branches hidden under the thick snow are heated by the spring sunshine and start to thaw rapidly. Although the roots are still buried in frozen ground, the branches begin to swell and produce flower buds, appearing in the form of furry catkins.

The mass appearance of various insects – mainly so-called 'snow fleas' (genus *Podura*) – on top of the snow and ice has for a long time been considered one of the most interesting phenomena of life in the Arctic.

These are small wingless insects belonging to the Collembola. They are found on top of snow at all latitudes but are especially common in polar areas and in high alpine environments. On snow on the tundra several scores of species of collembolans, staphylinid beetles and various Diptera belonging to the families Dolichopodidae and Ephydridrae may be found during the spring.

The winter-over insects are obviously heated during the spring by the sun's rays and crawl up to the surface. But it is not known how they succeed in penetrating the thick snow layer. They are often found on the surface in such great numbers that they completely cover the snow. Their quickest way up, is along thawed out plants, which reach above the snow. If during the spring, branches of willow bearing sprouting leaves or catkins, or the tall shoots of herbaceous plants, e.g., wormwood (*Artemisia*), are dug up from beneath the snow and shaken on to a sheet of paper it is possible to demonstrate that many small insects are alive under the snow. Collembolans and ichneumonid flies, etc., may be seen jumping about on the paper. The impression is that they were very close to the surface of the snow and, thus to the effects of the spring sunshine.

The collembolans found on the snow are very dark blue or violet–black in colour. The body is covered with thick hairs or by water-repellant scales. Therefore they may be found on melting snow, on ice or even on the surface of water, where they can even jump, being propelled by a flexing tail, the so-called 'spring tail' (see Fig. 24). Apparently the main purpose of their appearance on the surface of the snow is to utilize the heat of the spring

Fig. 24. The springtails most often encountered on the snow on the tundra. 1, *Ballistura*; 2, *Podura aquatica*; 3, *Hypogastrura*; 4, *Istomurus*; 5, *Isotoma*.

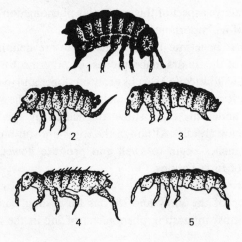

sunshine. Thanks to this, their period of active life is prolonged, improving their possibilities for completing their life cycle.

Collembolans, especially their oldest and most primitive groups, are in general characteristic of arctic and high alpine landscapes, where they occur in great numbers. It is possible that they are associated with snow, i.e., are chionophilous. This is a primitive biological characteristic of this particular group of insects. This kind of adaptation is no doubt useful in various types of landscape and may have developed just because of the polar and high alpine spring conditions or in connection with glacial periods of our planet.

The speed with which the snow disappears and its distribution over the tundra during spring are very important factors for nesting birds. For large birds of prey, e.g., falcons (*Falco peregrinus*), the rough-legged hawk (*Buteo lagopus*) and the snowy owl (*Nyctea scandiaca*) and also for geese (*Anser* spp., *Branta* spp.), the period between nest building and when the young start to fly may be as long as $3\frac{1}{2}$ months. Therefore the birds have to start nesting about the end of May or early June because by September cold weather returns, snow begins to fall and they must leave for the south.

At the end of May and the beginning of June snow covers the tundra more or less completely and there are few nesting sites. Some authors hold this fact responsible for the well known 'co-habitation' of birds of prey and 'peaceful' birds which may nest side by side, e.g., rough-legged hawks, peregrine falcons, and snowy owls with geese and black geese (or brants). It is said that during spring on the sites to have thawed first, birds of all kinds have to settle together. Therefore a protective attitude has also become necessary as a secondary characteristic of the feathered predators. Close to their nests they do not display their hunting instinct and they do not threaten the geese nearby. They will even protect not only their own nest against animals such as foxes but also at the same time protect those of the geese around them.

The slow emergence of the tundra from under the snow is of particular importance for the life-cycle of the snow geese (*Chen caerulescens*) nesting on Ostrov Vrangelya (see Fig. 25). Their colonies occupy strictly defined territories but at the same time the pairs will not nest too close to each other; each needs its own special territory. In order to raise their offspring successfully, the snow geese must start to build their nests there at the end of May or the beginning of June. In some years large parts of the tundra are free of snow at this time and the geese immediately distribute themselves throughout the nesting area. But often there is hardly any bare ground even at the beginning of June. Under such conditions there is keen competition for nesting sites. The geese constantly peck at each other.

When the nest is built, the pair defends it ferociously. Not all of them are able to find accommodation within the space available to the colony. Some pairs are left without a nest. The 'homeless' geese closely watch the more successful pairs and if they notice a nest left unguarded, they will occupy it and defend it as their own or quickly lay their eggs in it before flying away. As a result up to 20 eggs may be found in the same nest instead of the normal clutch of three to six eggs. However, the 'homeless' geese do not always succeed in laying mature eggs in a strange nest. Some of them are forced to lay them wherever they can. Usually the majority of such eggs are scattered around the territory.

The period when the snow melts may be a catastrophe for small animals such as lemmings, voles, shrews and sometimes even for stoats and weasels. At that time many of the young generation born during the winter perish when meltwater drowns those still in the nests. Nests inside the snow thaw out and are destroyed by the water. The summer burrows and tunnels made in the thick moss carpet during winter may previously have been plugged with ice and now may fill with water. The main occupation of the lemmings during the time of flooding is a move from 'winter quarters' to 'summer homes', i.e., from low-lying places and hillsides with a thick covering of snow to hill tops and watersheds in the moss-covered tundra. There the lemmings can dig new tunnels and nests or repair thawed-out old ones in the sward. Those animals who occupy burrows high on south-facing slopes with a light soil have the best sites. These habitats become free from snow early and dry out quickly.

Fig. 25. The blue or snow goose (*Chen caerulescens*) arrives on Ostrov Vrangelya while it is still snow covered; the pairs await the appearance of thawed-out patches (photo.: A. V. Krechmar).

A serious event in the life-cycle of plants and animals is the occurrence
of late frost and snow. In the southern tundra this is fairly rare, but in the
arctic subzone snow can fall at any time of the year. Spring and summer
snowfalls are harmful to flowers, the nests of birds and to some insects,
but especially to bumble-bees (*Bombus*) beginning to rear their offspring.

V. D. Aleksandrova (1961) described the effect of summer snowfalls on
the vegetation on the arctic tundra of Bol'shoy Lyakhovskiy Ostrov in
these words:

In 1956, snow fell there repeatedly during summer. Four of these snowfalls were
accompanied by freezing air temperatures. The first snowstorm occurred on 12 July
and the air temperature fell to -0.4 °C. Snow covered all the tundra but melted
quickly and did not cause any visible harm to the vegetation.

The second attack of frost occurred on 20 July. Towards evening on that date
snow began to fall. In the morning it covered the tundra with a layer up to 10 cm
thick, locally forming drifts. Shoots of fox-tail grass (*Alopecurus*), flowers of
poppies (*Papaver*) and buttercups (*Ranunculus*) stood up above the snow. On the
21st and 22nd the snow continued to fall intermittently. Only on 23 July did it finally
clear up. Toward evening that day the snow stopped entirely. The air temperature
had by then dropped to -1.5 °C. The majority of the plants had, however, not
suffered any harm. When the flowering buttercups became free from the icy crust,
they just continued to flower. The poppy flowers, on the other hand, had shrivelled
up, turned black and appeared damaged. On the 29th and 30th another blizzard
arrived and the air temperature dropped to about -0.7 °C. All the tundra was
covered by snow and the streams froze over.

When the snow stopped on the 31st, the poppy flowers had again perished. The
rest of the plant, however, continued living without noticeable change. The fourth
onslaught of cold and snow came on 5 August. At that time many new poppy flowers
had appeared, but for the third time that summer they were badly damaged by
frost. Snow fell on 6 August; on the 7th the temperature dropped to -2 °C. The
snow stopped on 8 August but the frost had of course finished off all the poppy
flowers. They stopped blossoming although there were still buds. For some days,
warm weather returned. But on the 10 August the leaves turned to autumn colours,
any remaining flowers quickly faded and the plants prepared to become dormant
for the winter.

This description gives an idea of how severe the conditions are in the
Arctic. There is actually a sort of 'constant spring' all summer long at very
high latitudes. The plants are remarkably well attuned to the sudden
changes in the weather. Only some of them, e.g., the poppies, are unable
to take freezing temperatures when flowering.

The snow on the tundra is very compact and therefore meltwater runnels
form on its surface. This so-called 'surface water' transports various plant
debris collected under the snow by the lemmings, such as remnants of food
and nest material. Meltwater can move this long distances and deposit it
on hill slopes and in ravines in rows reminiscent of hay raked together after
the harvest. Droppings of lemmings may collect along stream banks in

quantities of up to several kilos. The water carries away a great quantity of plant litter, especially leaves of birch and willow which also collect along river banks and on slopes of ravines. This is one of the reasons why so little leaf litter builds up on the tundra. In particular the accumulation of plant remnants on the surface of the moss carpet can change the character of the coenotic relationships and the direction taken by the soil-forming processes.

Water flowing from slowly melting snowdrifts plays a special role, and sometimes lasts into the autumn. On such sites peculiar snow-bed coenoses develop. Especially typical of these are the hygrophilous (water-loving) and cryophilic (cold-loving) mosses such as *Drepanocladus revolvens* and *Dicranum spadiceum*, among the shrubs the polar willow (*Salix polaris*), and among the herbs there is the butterbur (*Nardosmia frigida*) reminding us of 'coltsfoot' and some particular species of buttercup such as the tiny *Ranunculus nivalis* and *R. pygmaeus* as well as the large-flowered, fleshy-leaved *R. sulphureus*, etc.

In habitats where snow remains for a very long time a low-growing and patchy vegetation develops. Moss cover and shrubs are lacking, tiny herbs and dwarf shrubs predominate, e.g., buttercups (*Ranunculus*), saxifrages, mountain sorrel (*Oxyria digyna*), polar willow (*Salix polaris*), and so on. They grow very slowly; few of them flower or fruit. Their vegetative reproduction is also delayed. The plants look sickly and are not doing well.

The thermal regime of the tundra depends on the snow. During spring almost all the solar radiation is reflected back into space and utilized as thermal radiation or in melting. The reflective capacity of the snow depends on its content of dirt, its saturation with water etc. This results in a sharp increase in radiation balance during May. At that time the snow starts to melt, to form firn, baring debris accumulated below it. As a result the albedo diminishes rapidly. Snow melts faster around dark particles, forming deep pits. The surface of the snow expands in such a manner that it becomes porous. This speeds up its melting. At the point in time when the snow disappears completely, the solar radiation reaches its maximum but immediately begins to fall off again.

Thus, the different traits of the snow cover, its depth, duration, position, density, heat conducting capacity and many other factors have extremely varied and opposing effects on the soils, the vegetation and the animal life of the tundra. The thicker the snow cover and the more severe the winter conditions, the fewer are the species composing the vegetation and the animals colonizing the sites. However, different regimes, well adapted to severe arctic conditions, can produce a large amount of life. The more snow, the more fertile the soil and the richer the vegetation cover. It may result in nival meadow-like associations. However, in habitats with very thick snow drifts, e.g., at the foot of north-facing slopes where the snow melts extremely slowly, the ecological conditions become worse and the vegetation is impoverished. Only few species thrive, which are well adapted to the particular conditions existing there.

The Arctic is a nival landscape, a world of snow and ice. The duration of the snow cover is the main and the decisive factor in the lives of the majority of animals and plants. At the same time as the snow plays an important role, it also limits the opportunities for the existence of many species while still protecting them from the cold of the winters. By protecting the biotopes from winter frosts, the snow creates habitats for species of more southern origin. In areas where snow is scarce life is impoverished, but the process of selecting cold resistant varieties, well adapted to arctic conditions, is enhanced. All this contributes to the increase of a varied flora and fauna in the northlands and ensures the richness and the adaptability of tundra associations.

6

Adaptation of living organisms to conditions in the tundra zone

Examples of the marvellous adaptations of mammals, birds and many plants to the environment in the Arctic are of course well known to us from our school days. A wide range of scientific literature has been and is being devoted to such problems. At the same time, even now, very little is known about the general adaptation phenomena of organisms to severe conditions in polar lands. It is, for example, not clear whether the adaptive peculiarities inherent in all arctic animals and plants distinguish them into major groups and separate them from the inhabitants of other zones. In this chapter, we will not go into details but will concentrate our attention on the main ways in which organisms integrate with tundra landscapes.

Are living conditions in the tundra unique?

The natural environment of the tundra zone is extremely special. Therefore it is expedient to describe its main characteristics and to linger on the particular ways of assimilation of organisms into this landscape. Biologists try to understand the adaptive nature of arctic plants and animals, distinguishing them from the inhabitants of other zones. But this is not as simple as it may appear at first glance. In spite of the general peculiarity of the climate in the tundra zone, in addition some specific factors seem to be present in the environment which are uniquely related to each group of organisms.

Lack of heat and low temperatures are characteristic traits of living conditions in polar landscapes. At the same time minimum temperatures on the tundra are not extreme enough to exclude a large number of plants and animals because almost no organisms are subjected to the direct effect of frost during winter. At this time they are not active, they live mainly below the snow or have migrated to the south. The freezing temperatures in the tundra zone are, in reality, not as hard as in many other, more southerly continental areas (see Table 5). Many birds and mammals which survive belong to species which migrate from the tundra during winter such as the reindeer (*Rangifer tarandus*), the snowy owl (*Nyctea scandiaca*), the snow buntings (*Plectrophenax nivalis*) and ptarmigan and willow grouse (*Lagopus mutus, L. l. lagopus*). Just like typical taiga species such as the Siberian jay (*Perisoreus infaustus*), the Siberian nutcracker (*Nucifraga caryocatactes*), the titmice (*Parus*), squirrels (*Sciurus vulgaris*), wolverines (*Gulo gulo*) and sables (*Martes zibellina*) they have become adapted to the effects of hard frost. The inhabitants of the taiga are, actually, more cold hardy than the animals of the tundra. The plants on the tundra are covered by snow during winter but uncovered trees and the bushes of the taiga are able to endure extremely cold frosts. Indeed, there is no essential difference either in apparent cold hardiness of arctic invertebrates and those typical of more southerly landscapes (Lozina-Lozinsky, 1972).

One of the most characteristic traits of the polar environment is the almost continuous daylight during summer. However, for the life of the majority of organisms this is actually of greater importance in the forest tundra and in the northern taiga, especially in areas with a continental climate as, e.g., along the lower reaches of Khatanga, Anabar and Lena rivers where forested landscapes extend all the way to latitude 71–72 °N. There the climax of plant and animal life-cycles (in terms of the basic processes and in relation to reproduction) occurs during the brightest time of the year, in June and at the beginning of July. In the major part of the tundra zone 'summer time' does not start until the beginning and middle of July when the nights begin to darken again.

In Taymyr within the typical tundra subzone (at latitude 73 °N) in the middle of July when vegetative growth and flowering of plants reach a climax and the growth and reproduction of most animals is at a peak, the luminosity of the midnight sun is ten times less than that at noon. In relation to this, traits of diurnal rhythm have developed in the behaviour

Table 5. Air temperature in the tundra and the taiga regions of Siberia

Long term temperatures	Typical tundra: on the Pyasina lowland, Taymyr	Verkhoyansk	Yakutsk
Mean annual	−14·0	−15·7	−10·3
January mean	−35·0	−48·6	−43·2
July mean	+10·5	+15·2	+18·7
Minimum	−50·0	−67·8	−64·0
Maximum	+29·0	+34·6	+38·3

of the animals (see Fig. 26). By 8 August, the sun touches the horizon at this latitude. At latitude 68 °N, in the western sector of the subarctic where a southern tundra has developed, the arctic day ends on 19 July.

The most adverse factor for the living conditions in polar lands is actually the short time of darkness when processes take place which are necessary for the vegetative life of plants and the active life of poikilo-thermal (cold-blooded) animals and for the reproduction, growth and development of the majority of the homoiothermal (warm-blooded) animals. However, relatively, this factor is of concern only for some groups of animals, for others it is quite insignificant. It is more important that some plants and animals are able to remain active and to utilize the solar radiation even when covered by snow and at freezing temperatures. Indeed this makes it possible simultaneously to complete the life cycle and the reproductive processes during the short period of summer conditions. In the case of the lemmings, the optimum time for reproduction is winter. They raise their litters under snow. It is during winter reproduction that the level of their numbers is determined. In contrast, during spring and at the beginning of summer when water may fill the burrows and nests of the lemmings it is a very trying time for these little animals.

The brevity of the arctic summer presents great difficulties for the reproduction of many birds although not for all of them. Large species, especially predatory birds such as gyrfalcons (*Falco rusticolus*), rough-legged buzzard (*Buteo lagopus*) and snowy owls (*Nyctea scandiaca*) as well as the water birds such as geese (*Anser*), brants (*Branta*) and swans (*Cygnus*) indeed hardly have time to complete their breeding cycle during

Fig. 26. Daily course of: 1, daylight and 2, the intensity of flight activity of the bees in Taymyr at latitude 73 °N, 27 July 1969.

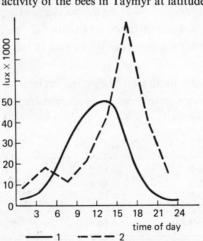

the short summer. Therefore they must lay their eggs as early as possible. They begin nesting even before the snow cover has disappeared completely. Often the eggs perish during late spells of frost and falls of snow. But for small passeriform birds, snipe and waders the short arctic summer is hardly of any major consequence. Thus the nesting period of small birds – from the construction of the nest until the young ones take to the wing – only lasts about 30 days. If the Lapland buntings (*Calcarius lapponicus*) lay their eggs at the beginning of July, the fledglings are fully grown and ready to take care of themselves by the middle of August while it is still warm and there is adequate food for them. However, a second brood such as many birds produce at middle latitudes is impossible under the conditions of the tundra zone.

Tundra plants not only are able to complete their life cycles but even to display seasonal aspects. Thus, in the typical tundra subzone the time of flowering can be distinguished as the early spring flowering of willow shrubs and the purple saxifrage (*Saxifraga oppositifolia*), the spring flowering of plants such as the blackish oxytrope (*Oxytropis nigrescens*) and Oeder's lousewort (*Pedicularis oederi*), the summer flowering of the great majority of the tundra herbs, and the late summer or autumn-flowering of plants such as arctic hedysarum (*Hedysarum arcticum*), Middendorff's larkspur (*Delphinium middendorffii*) and Tiles' saussurea (*Saussurea tilesii*). For instance, when the larkspurs are in full bloom, the blackish oxytrope is already in seed.

Thus, in spite of the general severity, the conditions in the tundra landscape provide acceptable habitats for species with different life styles and converging phenology where the organisms are able to survive in specific habitats although not everywhere in the landscape: on slopes or on flat land, among boulders, inside the moss cover or even in the soil. In the various habitats general climatic conditions undergo a greater or lesser transformation. On south-facing slopes suitable conditions may be found for many heat-loving species and close to snow beds there are conditions for cryophilic species.

In spite of the general poverty of organic life on the tundra, it harbours species displaying all kinds of morphology, physiology and behaviour. There are, however, also some unique and specifically adapted forms which are sharply distinguished from animals and plants inhabiting other zones.

Psychrophytes and cryophytes

Many arctic plants display characteristics typical of inhabitants of localities with hot or cold climates: their leaves are stiff with a hard epidermis, fleshy, needle-shaped or pressed flat to the stem (see Fig. 27).

Such peculiar traits are referred to under the term of xeromorphism. The ecological function is to reduce evaporation. Botanists distinguish two groups of tundra plants with such characteristics: psychrophytes and cryophytes. Psychrophytes are plants physiologically adapted to habitats in cold and wet conditions. Cryophytes are plants of cold and dry habitats. These two groups are almost equal in size. Frequently plants of the same species thrive both in dry and in very moist habitats. It is possible that all tundra plants with such a structure ought to be called psychrophytes because they very rarely occur under conditions of a truly physical drought.

Perennials, shrubs and dwarf shrubs such as avens (*Dryas*), arctic bell-heather (*Cassiope*) and some saxifrages (*Saxifraga oppositifolia*, *S. flagellaris*), crowberries (*Empetrum* spp.) and the penny-leaved willow (*Salix nummularia*), various grasses (*Deschampsia glauca*, *Koeleria asiatica* and *Festuca brachyphylla*), etc. are characteristically xeromorphic psychrophytes and cryophytes. The majority of these are widely distributed over the southern and typical tundra subzones but only a few such as *Saxifraga oppositifolia* and *S. flagellaris* are typical of the arctic subzone.

The occurrence of xeromorphism in habitats with high humidity is explained by the term 'physiological drought', i.e., the incapacity of the plants to absorb water under conditions of low temperature as a result of which physiological processes are slowed down and the suction capacity is diminished. At first glance it may appear that psychrophytes are plants typical of the tundra because the soils are almost always moist and cold there. However, very typical psychrophytes with clearly xeromorphically structured organs are characteristic more of the hyparctic belt (i.e., the southern tundra, the forest tundra, the northern taiga and the northernmost mountainous areas) than of the real tundra. Commonest in that belt are

Fig. 27. Tundra plant species with typical xeromorphic leaf structure. 1, Bell-heather (*Cassiope tetragona*); 2, purple saxifrage (*Saxifraga oppositifolia*); 3, spotted dryad (*Dryas punctata*); 4 flagellate saxifrage (*S. flagellaris*).

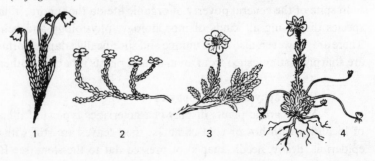

1   2   3   4

such plants as crowberries (*Empetrum*), *Andromeda*, Labrador tea (*Ledum* spp.) and cowberries (*Vaccinium vitis-idaea*).

Many typical psychrophytes are now represented by bog and mire species of *Andromeda, Vaccinium oxycoccus* (= *Oxycoccus* sp.) and Labrador tea (*Ledum*). The traits of xeromorphism are obvious in bog plants. There is of course much moisture there, but at the same time it is a cold biotope. The thick moss carpet and the peat are obstacles that prevent heating. The conditions on the level tundra are close to those of the mires. Therefore mire plants have penetrated out over the tundra. However, in addition, many other such species are absent from the arctic tundra subzone. Neither are there any of the species characteristic of drier biotopes such as the arctic bell-heather (*Cassiope*), the cowberry (*Vaccinium vitis-idaea*) or the penny-leaved willow (*Salix nummularia*) which are so common in the southern belt of the subarctic.

Conifers, the basic environment-forming species of the taiga, also represent a special kind of psychrophytism. Traits of xeromorphism are present in their leaves. The origin of these conifer characters is, however, not related to moisture. They are caused by the special system of water conduction typical of the gymnosperms, the absence of vessels, which impedes the percolation of water from the trunk to the leaves. For the northern conifers, however, xeromorphism plays a role of greatest importance during the severe winter conditions when drying out is especially perilous since at that time no water can flow to the leaves. There is evidence that xeromorphism may also develop under conditions of nitrogen shortage.

In the tundra, xeromorphism is to a great extent typical of species belonging to intrazonal biotopes, especially to those on dry and stony eroding slopes, on tops of hillocks and so on. In just such habitats the purple saxifrage (*Saxifraga oppositifolia*), *Dryas*, the arctic bell-heather (*Cassiope*) and the crowberry (*Empetrum*) are very common. They are considerably less common on level tundra.

A multitude of species of typical tundra plants lack any traits of xeromorphism. Most of these are legumes or bushy willows, many belong to the louseworts (*Pedicularis*), saxifrages (*Saxifraga*) and buttercups (*Ranunculus*), etc. Together with *Dryas* and *Cassiope* we find typical mesophytic and hygrophytic species such as golden saxifrages (*Chrysosplenium*), cloudberries (*Rubus chamaemorus*), dandelions (*Taraxacum*), mountain sorrel (*Oxyria digyna*) and species of milk vetch (*Astragalus*). On sandy and stony hilltops xerophytes with clearly visible xeromorphic traits dominate, i.e. *Dryas* and grasses like *Poa glauca* and *Hierochloë alpina*. In such habitats the physiological drought of the soils may be aggravated and periodically they may be physically dry. This can also be stated for the warm and gravelly soils of the arctic tundra of which, for instance, species of true cryophytes such as the saxifrages (*Saxifraga oppositifolia* and *S. flagellaris*) are typical.

As stated by the specialist on the arctic flora, B. A. Yurtsev, many tundra species, originating on the steppes, had already developed some xeromorphic traits there. Therefore tundra species such as wormwood (*Artemisia*), some xeromorphic sedges (*Carex*) and fescues (*Festuca*) are often no less xeromorphic than the closely related steppe species.

Thus xeromorphism in arctic plants is, apparently, more biotopic than zonal. In the present case this general rule is evident: problems of adaptation to some kinds of ecological factors are solved in different ways by different groups. Here, the physiological drought of the tundra soils is far more complex than revealed by the structure of the plants. In some cases in their habit they distinctly reflect a special kind of environment, in other cases the major role is played by other factors.

Is there any general trend in the adaptation of animals to tundra conditions?

Zoologists have often stated in their publications that strictly defined general traits of anatomy and physiology are peculiar to large groups of animals in the tundra zone, e.g., mammals and birds. These characteristics distinguish them quite sharply from animals of other landscape types. To such traits belong increased cold-hardiness, rapid development and fast growth and in general more intense life processes, in other words, all the particular aspects of life enhancing the anatomy of the organisms and allowing them to be less dependent on the capriciousness of the arctic climate. Attempts have long been made to fit all vertebrate animals there into a single anatomical and physiological type in full agreement with the conditions existing in the tundra landscape. Thus zoologists claim to have observed a single trend in the adaptation of organisms to conditions in the Arctic.

But these ideas appear convincing only at a general level of argument and speculation. When actual investigations are made, a multitude of factors crop up, contradicting these hypotheses. Thus, research on the adaptation of arctic and alpine vertebrates (e.g., by Shvarts, 1963; Danilov, 1966; Bol'shakov, 1972 also by Denisova and Artamonova) has revealed that tundra birds and mammals often have a lower rate of metabolism, a lower growth rate, a more slowly developing sexual maturity and a lower degree of fertility. This is particularly true of ancient and typical inhabitants of the tundra landscape. For instance, most representatives of a more southern fauna penetrating into the tundra may display characteristics such as a high intensity of their life processes, increased growth rate and rapid development as well as early maturity. Thanks to this they can adjust to the southern areas of the tundra zone but appear unable to cope in the habitats with more severe conditions in the northern areas.

Adaptation to the peculiarities of tundra life depends not only on climatic, ecological and biotopic conditions but also on characteristics of the biology and size of the organism, the history of the species in question, the level of phylogenetic development and the variability within a given group, etc. Each species encounters its own problems of adaptation to severe arctic conditions but often completely opposite types of adaptation develop under similar conditions.

Some groups of tundra fauna, due to their adaptive peculiarities, seem to refute the existence of any absolutely general principles of evolutionary adaptation. There is thus no doubt that the variability and the beautiful colouring of birds increases from high latitudes towards the Tropics. One of the reasons is the increasing richness of the fauna, because of which the importance of distinctive characteristics grows.

Snipe and waders present, however, a directly opposite example. The majority of these birds at middle and low latitudes are uniform in colour and drab, sporting predominantly grey, brown or ash-coloured hues. In the tundra zone, on the other hand, these birds are always variegated and brightly coloured. Such rusty brown tones with variegated designs and spots are typical of, e.g., the curlew sandpiper (*Calidris testacea*), the knot (*C. canutus*), the grey or red phalarope (*Phalaropus fulicarius*), the dotterel (*Eudromias morinellus*), the turnstone (*Arenaria interpres*), the black-bellied plover (*Pluvialis squatarola*), the golden plovers (*Pluvialis dominica* and *P. apricaria*) and the ruff (*Philomachus pugnax*), etc. This is an example of the great diversity and abundance of wading birds and snipe within a single habitat in the tundra landscape. Here the role of distinctive characteristics is increased just as in the case of many birds in the Tropics.

Another fine example of adaptive variation to tundra conditions, and to specific aspects there, is found among the bumble-bees (*Bombus*). As demonstrated in the classical works by Strel'nikov (1940) bumble-bees are a special genus of 'warm-blooded' insects. Thanks to their big and densely hairy bodies they are able to conserve heat produced when flying. Therefore they can be active at very low temperatures. It must of course be pointed out that the large size of the bumble-bee, ideal for more efficiently preserving heat (because of the smaller relative surface of the body), is a useful characteristic under arctic conditions. There are actually

very many large species of bumble-bees in the tundra zone, such as
B. hyperboreus, *B. pyrrhopygus*, etc.

An inevitable consequence of large size is a prolonged time of development. In this connection the reduced number or the complete absence of
worker bees is of great importance for the arctic bumble-bees. Due to this
the queen herself must raise both females and males. However, one of the
smaller-sized species, *B. lapponicus*, is also widely distributed within the
tundra. It has the normal number of workers. Of course, the development
of a small-sized bumble-bee is considerably faster than that of a large-sized
species. Therefore they can successfully raise workers as well under the
conditions of the short arctic summer. Thus, small size, while not very
favourable from the point of view of adaptation to cold, nevertheless
permits polymorphism to be preserved within the bumble-bee family under
severe arctic conditions. Species with large bodies, advantageous in terms
of thermoregulation, have been forced to sacrifice their workers.

This example once more demonstrates the fact that adaptation to
conditions within the tundra zone often takes opposite directions. Small
size or large, slower or more rapid development depending on various
circumstances can all play positive roles in the basically severe environment
of the subarctic.

Thus, in the tundra zone there are most varied adaptations: different
organisms adapt in opposite ways. The subarctic conditions may be
extreme but they are also sufficiently variable to provide habitats for
different kinds of life forms always giving rise to well saturated and
complete associations. In the adaptations of the tundra organisms we can
also see reflections of the landscapes of the past, of the history of the
species, of the routes leading to colonization of the subarctic as well as of
the differentiating effects of the arctic environment and of specific
characteristics of the species themselves, their physiology, their habitats
and their interactions with other organisms.

The role of the history of the species

Characteristics not developed in the subarctic may, under different
conditions, often prove especially useful in the tundra zone. A fine example
is the subnival reproduction of tundra rodents such as lemmings and
narrow-skulled voles. Shvarts (1963) stated that subnival reproduction is
a kind of adaptation making it possible for these rodents to cope
successfully with tundra conditions and to allow them to produce valuable
and numerous offspring independently of the capriciousness of the arctic
summer. This biological trait developed, however, not solely as a result
of adaptation to tundra conditions.

The narrow-skulled vole (*Microtus gregalis*) originated in the steppe zone. Another subspecies still exists there which also reproduced under the snow during winter. This is typical also of other steppe rodents. Having colonized the tundra the narrow-skulled vole developed into a different subspecies for which subnival reproduction proved a very useful adaptation under existing conditions. The ancestors of the tundra lemmings also came from the steppes. The closest relation of the collared lemming (*Dicrostonyx torquatus*; see Fig. 28), during the last glaciation, was a common inhabitant of the so-called periglacial steppes in the area extending from Central Europe to Eastern Siberia.

Many such cases are common in the flora and fauna of the subarctic where species have not really adapted to arctic conditions. It appears rather as if the adaptation had already developed in response to the demands of a given type of environment. This is called pre-adaptation. Simply stated it is the predisposition of an organism for this or that type of condition. It applies in equal measure both to separate species as well as to whole groups of animals and plants. Thus, the astonishing adaptation of mosses and lichens to the severe arctic conditions is well known. But basically nothing is stated regarding the fact that the extraordinary cold-hardiness of mosses and lichens could be the result of any spontaneous adaptation to severely cold climates. The main characteristic of the flourishing mosses and lichens is their generally low metabolism, their ability to stop growing at any time and to resume growth without any difficulty when conditions improve. These capacities can be utilized as an adaptation to hot climates as well.

Fig. 28. The collared lemming (*Dicrostonyx torquatus*), a descendant of the periglacial steppe rodents (photo.: A. V. Krechmar).

It is possible that the decisive factors here are the generally not very high phylogenetic level of the group, their obviously primitive state and their capacity for low-intensity life processes.

Shvetsova and Voznesenskiy (1970) were unable to discover any specific peculiarities regarding the photosynthesis of tundra plants. There are species among them, just as in the forests, with both low and high intensity photosynthesis. Most important of all is the fact that the upper temperature limit for photosynthesis in some tundra plants is 50 °C. Such high temperatures do not occur in the subarctic. It is possible that this trait represents a case where plants retain an adaptation developed under conditions of other and considerably warmer landscapes. In the case in question this characteristic is not of any importance for life in the tundra and could be considered as an atavism. Similarly the peculiarities of the daily rhythm are of interest. In polar lands a diurnal rhythm has developed in some widely distributed species similar to that of much more southerly populations existing under quite different light conditions (see Fig. 29).

Such characteristics may occur in species inhabiting the subarctic which, so to say, are not typical of tundra conditions exclusively. Thus hibernation or aestivation is characteristic of such native inhabitants of arid landscapes as gophers, marmots and jerboas. For inhabitants of high latitudes this type of adaptation cannot be widely applied because it is very difficult during the short summer to build up adequate fat reserves necessary for the long winter. However, one of the hibernating species which has penetrated into arctic landscapes and inhabits the tundra in Chukotka, Alaska and Canada is the long tailed American arctic ground squirrel (*Citellus parryi*; see Fig. 30), closely related to the narrow tailed Asiatic

Fig. 29. Variation in vocalizing intensity of common wheatear (*Oenanthe oenanthe*): 1, in western Taymyr; 2, near Voronezh, at latitude 52 °N (after Podaruyeva).

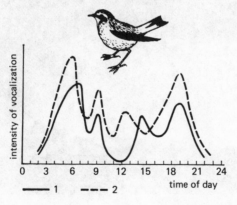

ground squirrel *C. undulatus*. The latter animals are natives of steppes and alpine meadows.

The landscapes of Chukotka where the American arctic ground squirrel lives represent a special variant of southern tundra. There this animal has selected herbaceous and meadow-like habitats along river banks, on slopes of ridges and hills and among block scree. This large rodent needs an abundance of juicy, green fodder. It is exactly its ability to reduce its active metabolism during winter that makes it possible for it to colonize the tundra. However, the fat reserves accumulated during the summer are barely adequate for the period of hibernation during the long arctic winter. Therefore these ground squirrels start hibernating very late in the autumn. They remain active during strong frosts until thick snow has fallen. In the spring they appear early on the surface of the compacted snow cover and they make deep trails and paths in it when searching for food.

However, many species inhabiting the tundra zone are actually pre-adapted to severe conditions. The particular biological trait useful when coping with tundra conditions may, of course, in the long run change or be reinforced when the taxon in question becomes a specialized inhabitant of the arctic environment.

Importance of body size
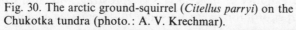
The smaller the organism, the more easily will it find suitable living conditions. Unicellular algae can inhabit lakes, puddles and films of water

Fig. 30. The arctic ground-squirrel (*Citellus parryi*) on the Chukotka tundra (photo.: A. V. Krechmar).

surrounding soil particles, but for larger aquatic animals certain types of permanent water bodies are necessary. Among the microscopic tundra organisms there are also many species distributed in forest steppes and even in deserts. These are various kinds of soil mites, collembolans and nematodes. These small organisms are often not especially adapted to the habitats in the tundra. Although these may be taxa characteristic of the tundra and be met with exclusively within this territory, they are very similar to relatives in landscapes further south.

Small animals such as lemmings and stoats spend the winter under the snow and are not really subjected to hard frosts, but the northern reindeer and the foxes cannot hide from the cold temperatures during winter and must be able to cope with arctic frosts.

Above we discussed the importance of snow abrasion of trees and bushes. This phenomenon is, however, of no importance to the overwhelming majority of herbs and shrubs because they are covered with snow during winter. Thus, the larger the organism, the more it is subjected to the direct effects of the climatic conditions in the tundra and the greater must be its capacity for coping there.

The low stature of tundra plants, their 'dwarfism', is one of their most noticeable external characteristics. Individual tree species, extending from more southerly areas, grow low and often form krummholz, their stems and crooked branches pressed flat to the ground. The height of the tundra shrubs is often only 0.5–1 m. Dwarf birches and willows creep among the mosses. The majority of their species are prostrate with stems buried in the moss carpet or jutting above it some 10–15 cm as is common in, e.g., the arctic willow (*Salix arctica*) and the 'beautiful' willow (*S. pulchra*).

Still smaller dimensions are found among some shrubs now dominating certain tundra environments. These are also well known inhabitants of temperate climates such as cowberries (*Vaccinium vitis-idaea*) and bog whortleberries (*V. uliginosum*) as well as such 'tundra inhabitants' as the arctic bell-heather (*Cassiope*), the dryads (*Dryas punctata*, *D. octopetala*) and many others. Usually they are about 10 cm tall but often only 3–5 cm. The smallest dimensions of all are typical of one of the tiniest shrubs in the world, the polar willow (*Salix polaris*). The thin stem of this taxon is buried in the moss carpet and has only two or three leaves, *c*. 1 cm in diameter or less, penetrating above the mosses. The height of herbs, grasses, sedges and so on average 10–15 cm and only the reproductive shoots attain a height of up to 20 cm at the time of fruit dispersal.

One of the most typical peculiarities of the development of tundra plants is the fact that they flower on extremely short pedicels. Many plants flower when the ground is barely free of snow. At that time, growth has hardly started and the peduncles are very short. This phenomenon is, of course, of adaptive importance. The air temperature is highest in the layer closest to the ground, thus making it possible for the plants to flower during the generally cool temperatures during spring. For the dispersal of seeds, which for most plants is assisted by winds, taller peduncles are expedient. The stalks of spring flowering plants grow four to five times taller after flowering, but those of summer flowering taxa may elongate during bud formation. Arctic poppies (e.g., *Papaver polare*) have rather tall peduncles, 8–12 cm. However, in the arctic tundra, these are often prostrate and slightly winding over the ground surface where, in comparison with the air temperatures higher up the conditions are more favourable for flower development.

Together with the small dimensions of tundra plants the remarkably large size and bright colours of their flowers are typical. Arctic flowers as such are not larger than those in the forests but considering the dimensions of the peduncles and the leaves they seem to be rather large. Just because of this, tundra plants are a favourite subject of photographers. Extensive areas can be photographed showing plants which have an abundance of large flowers with diameters up to 3 cm, such as the arctic avens (*Novosieversia glacialis*), the pulvinate poppy (*Papaver pulvinatum*) or the large-flowered chickweed (*Cerastium maximum*). Often many small flowers form a large inflorescence which makes them more spectacular as, for example, in the milk vetches and the louseworts (*Astragalus, Pedicularis*).

The low growth of plants is associated with the general severity of the climate such as a lack of heat, more favourable temperature conditions in the layer of air close to the ground, scanty snow cover, lack of nitrogen, etc. Tall growth is correlated with great expenditure of nutrient matter. Low growth promotes economy of the nutrition available which is important for the ripening of seeds and for completing the life-cycle.

The dwarfism of tundra populations of many species is of course genetically determined. This has been demonstrated experimentally by American scientists on some small plants belonging to the family Cruciferae, i.e., the alpine cress (*Cardamine bellidifolia*). Seeds of these plants were collected in an alpine meadow in the mountains of Alaska and on the islands of the Canadian Arctic Archipelago at latitudes 80–82 °N. When raised under identical temperate temperature conditions, the alpine plants grew taller but the arctic ones remained as they were in the tundra, small but with a large number of tiny leaves.

Small body size is also advantageous for arctic invertebrates. The smaller the animal, the easier it finds suitable living conditions and the faster it completes growth and development. In this case there are considerably

larger numbers of small-sized species of tundra insects and other invertebrates than in the fauna of more southerly zones. This is especially
distinctive in the polar desert where there are no large insects such as
butterflies, carabids, bumble-bees or water-beetles, etc. There, only small
representatives remain, collembolans, chironomids, snow and fungus flies,
small moths and some ichneumonid flies. In general, however, the process
of 'reductionism' at high latitudes is less common in animals than in plants.
In the arctic tundra subzone many large insects are still common such as
bumble-bees, craneflies and some Diptera, e.g., deer flies and hoverflies.
Besides, the craneflies or tipulids – the largest members of the mosquito
group – belong to one of the largest faunal groups within this subzone.

The reduction of body size among warm-blooded animals is very
complicated at high latitudes. According to the well-known 'Bergman
rule', established already during the last century, the body size of many
taxa among the native species or subspecies of homoiothermal (warm-
blooded) animals will increase from the equator towards the polar areas
(see Table 6). The larger the body, the smaller is its relative surface and,
consequently, the less heat it radiates. In other words, the larger the animal,
the more easily it retains heat. Bergman formulated his rule very cautiously
as a general tendency with a large number of exceptions. He understood
the great complexity of the problems affecting it and emphasized that body
size is just one of a multitude of adaptive factors typical of this or that
kind of climate, including the arctic. Based on a number of examples he
showed that the problem of thermoregulation is solved by each species in
its own way. In addition to body size, important roles are also played by
a furry or feathered covering, hibernation, type of shelter, etc.

Table 6. Northern limits of distribution and body sizes of bears*

Species	Approx. northern limit of area, °N	Maximum body length, m	Maximum body weight, kg
Polar bear (*Ursus maritimus*)	85	2·8	700
Brown bear (*U. arctos*)	72	2·4	500
Black (Tibetan) bear (*U. tibetanus*)	52	1·7	150
Malayan bear (*Melarctos malayanus*)	25	1·4	65

* From V. G. Heptner and N. P. Naumov (Eds). *Mammals of the Soviet
Union*, 'Vysskaya Shkola' Publishing House, Moscow 1967.

Previously some authors applied Bergman's rule in a very categorical manner. Thus, the American scientist Allen wrote: 'For instance, the dimensions of mammals and birds in North America increase from the south toward the north. This is apparent not only with respect to individuals of a given species but also in general regarding the largest species of every genus or family appearing farther northwards.' In the literature examples of larger sized species or subspecies are often pointed out. However, there are a multitude of cases in direct opposition to this rule.

There is no doubt that warm-blooded animals tend to have large bodies in cold climates. However, the zonal change in dimensions of birds and mammals is extremely complex and contradictory. When comparing the dimensions of birds living at high and very low latitudes, it seems that a tendency to reduction in body size of arctic species is definitely more consistent than of those native to the south. The ptarmigan (*Lagopus mutus*) extending farther north is smaller than the willow grouse (*L. l. lagopus*). The southernmost (forest–steppe) subspecies of the willow grouse is the largest. Bewick's swan (*Cygnus bewickii*) is the smallest representative of its genus; the smallest species of geese are arctic, e.g. the

Table 7. Indices of maximum weight and duration of development of some species of swans and geese (Syroyechkovskiy, 1978)

Species	Weight (kg)	Duration of incubation days (approx.)	Duration of development of nestlings, days (approx.)	Total days (approx.) required for development
Mute swan (*C. olor*)	16·0	36	150	186
Whooper swan (*Cygnus cygnus*)	13·0	40	120	160
Bewick's swan (*C. bewickii*)	6·0	30	45	75
Bean goose (*Anser fabalis*)	3·7	29	45	74
Blue (snow) goose (*Chen caerulescens*)	2·9	26	49	75
Lesser white-fronted goose *Anser erythropus*	2·5	28	40	68
Red-breasted goose (*B. ruficollis*)	1·7	26	40	66
Brent goose (*B. bernicla*)	1·4	26	40	66

brent goose (*Branta bernicla*), the red-breasted goose (*B. ruficollis*) and the lesser white-fronted goose (*Anser erythropus*; see Table 7). The largest birds in the tundra zone in general colonize its southern belt (see Figs. 31, 32).

In Shvarts (1963) the following information on the dimensions of tundra animals appears: of 20 species of mammals distributed both in the subarctic as well as beyond its limits, the largest dimensions within

Fig. 31. Below: a, whooper swan (*Cygnus cygnus*) and b, black-throated diver (*Gavia arctica*). Opposite: c, sandhill crane (*Grus canadensis*): the largest birds on the tundra (photo.: A. V. Krechmar).

predominantly arctic populations are found in nine species; lesser dimensions (by comparison with forest populations) in five species; three species are no different in size from the southern populations; for the rest, data are lacking. Irrespective of zonal changes in climate the body size of some species of mammals may diminish or increase in all directions away from a particular centre. Thus, the average length of the skull of the snow hare (*Lepus timidus*) is smallest in eastern Europe and western Siberia. It increases in size not only towards the north but also towards the east, west and south of that area. The smallest subspecies of the arctic hare exists under the most severe conditions in northern Siberia. Such examples are worth mentioning as is the fact that the body size of stoats (*Mustela erminea*) is large but that of some weasels (e.g., *M. nivalis*) is small. The tundra subspecies of the latter (the 'small' weasel, see Fig. 33) is, as its name implies, quite small. The tundra wolf (*Canis lupus*) is, on the other hand, considerably larger than the subspecies of the forest zone.

In spite of the obvious advantages of a large body, why are the dimensions of many warm-blooded animals at high latitudes so small? The main reason for this is the necessity of completing the reproductive cycle as fast as possible and of raising offspring during the course of a single summer season. These facts are most distinctly evident among birds, which

must be able to complete their growth and fly south before winter. The smaller the body size, the shorter is the period required for growth and development (see Table 7). For large species of birds, the period of nesting and maturation of the nestlings is compatible with the duration of the warm season in the tundra, whether short or long. The whooper swan (*Cygnus cygnus*) cannot survive in the arctic tundra subzone just because of the fact that there is simply not sufficient time of relatively frost-free weather for it to raise its offspring irrespective of other conditions. But also

Fig. 32. Above: brent goose (*Branta bernicla*); below: red-breasted goose (*B. ruficollis*), the smallest of the tundra geese (photo.: A. V. Krechmar).

the geese barely succeed in raising their young and therefore they must start nesting very early while the weather is still quite unfavourable. In some cases also the supply of food is, no doubt, of definite importance as well as the opportunities for locating enough to eat which can be a greater problem for a large-size organism.

The short arctic summer offers, indeed, many reasons for the predominance of small-size forms of insects and other invertebrates at high latitudes. True enough, the picture is somewhat different here than for larger animals. One of the obvious pathways of adaptation to the severe conditions in the tundra is the capacity for a delayed development, fully tested by the weather conditions. Many insects in the tundra develop over two, three and up to five seasons. Thus, at certain stages, they are not dependent on the short arctic summer or the capriciousness of the weather. But also in such cases small dimensions are an advantage because development of large-size invertebrates under the far more severe conditions in the arctic tundra and the polar desert may require an almost indefinitely long period of time.

Economical principles of body surface

Another of the main 'rules' of geographically important characteristics of warm-blooded animals is also well known, that is that concerning the reduction in relative size of protruding body parts in correlation with cold climates. This rule is called 'Allen's axiom'. It

Fig. 33. The smallest predatory animal to have extended into the tundra, the weasel (*Mustela nivalis*); (photo.: A. V. Krechmar).

concerns problems of thermoregulation. The more parts there are protruding from an animal body and the larger they are (e.g., ears, tails, legs), the greater is their heat radiation and the stronger the cooling of the organism. A lesser size of these organs is advantageous during cold conditions, a larger size in hot climates. Usually this axiom is illustrated by examples from some species of foxes. The shortest ears are found on the arctic fox (*Alopex lagopus*), those of the red fox (*Vulpes vulpes*) are somewhat larger and those of the steppe foxes are longer still and the very largest belong to the fennec of the African deserts.

This rule constitutes a particular case of very general importance which can be called the 'economical principle of body surface'. Under cold conditions the most advantageous form for a body is actually a sphere, when the ratio of surface to volume is at a minimum. Consequently it radiates less heat. Therefore the warm-blooded inhabitants of the polar areas have in addition to various adaptations for the preservation of heat in the form of fur and feathers also more compactly rounded body shapes with a minimum of protruding organs (see Fig. 34).

The importance of surface economy is to a great extent also evident in tundra plants. Many species have a basal mass of leaves collected into a rosette or 'cushion' just above the roots. Mosses and lichens often grow in the shape of dense, rounded clumps. The farther toward the north a zone is situated, the more distinct are these growth forms.

Dense leaf rosettes are typical of poppies (*Papaver* spp.), *Draba* spp.,

Fig. 34. The Siberian lemming (*Lemmus sibiricus*), a typical representative of the tundra mammals (photo.: A. V. Krechmar).

Saxifraga spp., arctic avens (*Novosieversia glacialis*) and many others. The more severe the conditions, the more noticeable is this growth form. For instance, specimens of *Dryas* growing in the moss cover have longer stems and a more open arrangement of the leaves than is common for this species on gravelly sites which are snow-free during winter. There they may form

Fig. 35. Dense-cushion of the tufted saxifrage, *Saxifraga cespitosa* (photo.: N. V. Matveyeva).

Fig. 36. 'Stripes' and 'cushions' of saxifrages, poppies and forget-me-nots on the arctic tundra.

10 cm

densely compacted stands with branches thickly covered by leaves. The woolly forget-me-not, (*Eritrichum villosum*), in the southern tundra zone forms very loose tussocks, breaking up into individual branches. In the north it is almost spherical in shape. Many species of saxifrage (e.g., *Saxifraga hirculus, S. cernua* and *S. cespitosa*) are many-stemmed and have small leaf rosettes in the southern and the typical tundra subzones, while in the arctic tundra they are many-stemmed but form a dense clump (see Figs. 35, 36). The largest and densest cushions are formed by the arctic avens (*Novosieversia glacialis*) (see Fig. 37). Some specimens of these reach 60–80 cm in diameter and up to 15 cm in height.

Some species look like sprawling mats pressed flat to the ground. Such mats are formed in, e.g., the arctic tundra subzone by the sandwort (*Minuartia macrocarpa*). However, in the southern tundra zone this species has numerous loosely-arranged branches. Strongly reduced forms of early-flowering species are represented by the purple saxifrage (*Saxifraga oppositifolia*). In the southern parts of the tundra this herbaceous species has fleshy stems with the leaves situated at about 1 cm distance from each other. In the north or in alpine areas, naked woody stems with woolly-

Fig. 37. Well established tussocks, formed by the branches of the arctic aven, *Novosieversia glacialis*.

haired, leafy branches develop, forming dense cushions. Many such examples could be mentioned.

What is the ecological importance of such forms? Evidently bud development is better protected during autumn and winter against both snow abrasion and wilting. Perhaps it may be thought that the buds of arctic plants are better protected by hard bud scales, shielding them from mechanical damage and desiccation? But very often the buds are covered only by old leaves and sometimes not protected at all. Apparently hard bud scales are an obstacle to fast growth during the spring and, consequently, not advantageous to a plant under these conditions during the short vegetative period in the Arctic. When positioned within dense cushions or rosettes, growing points are safely covered below old leaves which although frozen will persist for many years on the plant. Within such cushions there are more favourable heat conditions during spring when the snow is just beginning to disappear.

The well known scientist B. A. Tikhomirov (1963) wrote about this phenomenon: 'the frozen and wilted leaves of *Novosieversia glacialis*, persisting on the plants for a long time serve, under unfavourable conditions, as a protection to rhizomes and buds during their renewal . . . Rising 8–10 cm above the ground surface, the branches of the arctic avens are, thanks to the abundant remnants of old leaves with holes in them, heated up by the solar rays when the snow cover is still present.' The formation by arctic plants of rosettes and cushions as well as their low growth forms enables them to function within the ground layer of the atmosphere which is more favourable for life during the summer.

Pathways leading to the formation of the tundra flora and fauna

The tundra and the polar desert are the youngest of the zonal landscapes on earth. During the major part of the Tertiary era, broad-leaved and coniferous forests developed in the area which is now the subarctic. Tundra vegetation started to form at the northern extremes of Eurasia and America in connection with the cooling of the climate during the time of the Quaternary glaciations. Paleobotany and paleozoology inform us of these events: that even at the start of the Quaternary period a typical tundra landscape existed with characteristic phytocoenoses and appropriate adaptations of the flora and the fauna to the conditions there. This causes us to assume that the nucleus of the arctic flora and fauna, or at least of their nearest precursors, had been formed before the glaciation and that they were able to adjust quickly to the new type of landscape. The majority of botanists believe that the main precursors of the tundra flora were plants existing under the cover of forests in the taiga zone but also in the alpine belts of high mountains. These species are believed to constitute the source of the tundra flora in the northern parts of Eurasia and America.

Even during the height of maximum glaciation, extensive areas in the northern parts of Asia and North America remained free of glaciers. This was primarily the case for areas in Eastern Siberia and Alaska. They acted as kinds of 'floral and faunal laboratories', supplying species able to form associations typical of the tundra. They also played a role in the origin and the organization of present life on the tundra. The ancient flora and fauna of these territories became modified, migrating away from the glaciated areas and becoming assimilated into a new type of landscape. In this connection an important role was played and is still being played up to this day by the colonization of the subarctic landscapes with more southern species, including cosmopolitans. The processes of repopulating the tundra zone and the formation of associations of animal and plant species have not ended even today. Wide areas are being opened up to new forms or compositions of arctic associations.

Expansion beyond their previous limits occurred primarily in those species which are in general widely distributed and inhabitants of several zones. Species which are common within the present tundra, such as stoats (*Mustela erminea*), willow grouse (*Lagopus l. lagopus*), the peregrine falcons (*Falco peregrinus*) and seven-spotted ladybirds (*Coccinella septempunctata*), may serve as examples. Such species are called polyzonal. On the other hand, species typical of the northern forest landscapes colonized the subarctic to a lesser degree. The species which were already characteristic representatives of the southern tundra fauna, the so-called hyparctic species, rarely expanded into more northerly latitudes. Their distribution is often sharply curtailed at the border of the typical tundra subzone. Thus the polyzonal snipe (*Gallinago gallinago*) and the stoat (*Mustela erminea*) are very widely distributed within the subarctic as far as its northernmost limit, while the pin-tailed snipe (*Gallinago stenura*) is characteristic of the forest tundra. Middendorff's vole (*Microtus middendorffii*) is common in the southern tundra but absent from the typical subzone. Paradoxical situations have developed: the typical inhabitants of the arctic landscapes colonize the tundra less intensively than many of the southern forms or even some secondary settlers of the subarctic.

This may be explained as follows. Adaptation to zonal conditions is always determined by the degree of specialization that make it possible to inhabit other zones. The more severe the climate the more comprehensive it is. It could be said that species of the forest steppe or of the forest are less specialized than those of the desert and that the tundra species are less adaptable than those of the polar desert. Therefore the species formed in a temperate climate are often more capable of expanding into the tundra zone than tundra species are of settling in forested areas. The typical inhabitants of the polar desert have the least capacity for expansion.

All this is reasonable not only with respect to the various zones but also to parts thereof. Many species of the temperate zone are more widely distributed within the forest belt than are arctic species within the tundra. A considerable number of species are well adapted to an existence within one single or at most two contiguous subzones of the tundra. The reason for this is the sharp change in conditions within the subarctic territory. The southern, the typical and the arctic subzones are sharply distinct from each other in particulars concerning climate and landscape type as can be seen clearly from the examples given of the variation in mean July temperatures (see Fig. 3).

The tundra is a young landscape but many of its inhabitants are species that have undergone extreme adaptation to the severe arctic conditions and have become isolated in special genera without close relationships. Examples are the Lapland bunting (*Calcarius lapponicus*), the snowy owl (*Nyctea scandiaca*), the reindeer (*Rangifer tarandus*) and among plants arctic avens (*Novosieversia glacialis*). These species are evidence of the fact that they have developed a long way under tundra conditions or in similar landscapes. Judging by the present representatives it is impossible to explain anything about the speed of their evolution or whether they were formed only after the tundra had developed as a natural zone. Most likely they existed much earlier. Again we encounter a paradox: the typical tundra species may be more ancient than the tundra landscape. Obviously the colonizers of the tundra were somehow pre-adapted to such conditions. In other words, they may have lived for long in biotopes similar to those of the tundra. Under the conditions of a milder climate, reigning on Earth during the pre-Quaternary time, e.g., during the Pliocene, their habitats may have been found exclusively on high mountains. Actually, among the typical inhabitants of the tundra, there are many descendants of alpine animals and plants. Because of this, very often the tundra populations have only subspecific rank. For example, species such as the dotterel (*Eudromius morinellus*), the ptarmigan (*Lagopus mutus*) and the liliaceous species *Lloydia serotina*.

Here again we meet with strict specialization during the evolutionary process and the adaptation to the conditions in the north within various groups of organisms. Arctic–alpine species occur especially frequently among forms of flowering plants and within the class of insects. Also included in the category are about six species of birds but almost no mammals. Shvarts (1963) stated regarding this problem that: 'one of the most important pathways towards formation of a tundra fauna (on the basis of alpine elements) for some reason proved inaccessible to the mammals'. None of the mountain species of mammals distributed within the subarctic has wandered from the mountains down onto the plains, nor have they become typical tundra animals or part of the present tundra zonal biocoenoses. They have all remained on the high mountains. This concerns such wild animals as the mountain sheep, the marmots (*Marmota camschatica*), the pika (*Ochotona alpina*), the northern Siberian vole (*Microtus hyperboreus*) and the large-toothed red-backed vole (*Clethrionomys rufocanus*). The possibility exists that the determining factor here is that the tundra landscape is able to assimilate in general only a limited number of mammal species. The principal reason may be found in the main level of development and behavioural specialization of the organisms. The mammals with their advanced and complicated life forms and behaviour in the high mountains and in the tundra basically all belong to different biotopes in spite of their apparent resemblance. Therefore species which are adapted to the conditions of alpine belts may be unable to move into the tundra.

Many representatives of animal groups living in the tundra may seem incompatible with this type of landscape, such as steppe species. Some species from the steppes are, for example, the salchil bug (*Chiloxanthus pilosus*), the horned lark (*Eremophila alpestris*) and the narrow-skulled vole (*Microtus gregalis*). It is obvious that at high latitudes they have acquired some peculiarities in their biology, morphology, coloration and dimensions. They are usually classified as different subspecies. In other cases they are separate species but evidently descendants of 'steppe inhabitants'. Thus, a typical steppe species like the sagebrush vole (*Lagurus lagurus*) is closely related to the collared lemming (*Dicrostonyx torquatus*) in the tundra. Being immigrants from the steppes, they have adapted to tundra conditions thanks to various biological traits. In the case of the vole, this entails a subnival form of life, for the horned lark complex adaptations to habitats in types of landscapes covered by herbs as well as to polytrophy, i.e., the ability to utilize more than one kind of nutrients.

At present the climates of tundra and steppe are sharply different and the similarity of the species representing their faunas at first glance seems strange. But this problem has a special history. Even during the last century, studies of fossil remains of animals from the Pleistocene, that is, the youngest geological epoch on Earth, led to the discovery of peculiar phenomena. In different areas of northern Eurasia remains have been found together, of animals which at present have no connection. Some of these now inhabit tundra territories, others steppe or semi-desert areas.

Even forest species were met with among them. This peculiar fossil fauna was called a 'mixed type'. It included such animals as the steppe antelope (*Saiga tatarica*), musk ox (*Ovibos moschatus*), lemmings (*Lemmus, Dicrostonyx*), sagebrush vole (*Lagurus lagurus*), reindeer (*Rangifer tarandus*), wild horses (*Equus*), foxes (*Alopex, Vulpes*), etc.; for details regarding these problems, see the book by Vangengeym (1977).

Various hypotheses have been formulated to explain this phenomenon. One investigation states that the reason for it is to be found in seasonal migration of the mammals; another refers to the more even climate at that time as a result of which the animals had, in general, wider distribution areas. However, with the appearance of the sharply defined zonation of present types, they would have 'dispersed' into different zones. In modern times, more substantial points of view have been presented. Some were expressed even in 1928 by the eminent ornithologist A. Ya. Tugarinov who explained the mixture of arctic and steppe species as follows: 'the climatic and ecological conditions were such that they could satisfy to an equal extent the living requirements of both northern and southern inhabitants'.

Actually, according to data from paleoclimatologists, dry and cold air masses dominated the major part of northern Eurasia during the glacial period. Under their influence the so-called periglacial (referring to the region adjacent to an ice sheet) landscapes were formed, especially a kind of tundra steppe, combining the traits of both tundras and steppes. The climate was on the whole cold, but with hot summers. The lack of precipitation made the physiological drought more severe for the plants. In this peculiar landscape both cold-hardy and drought-resistant species, or species with various combinations of characteristics, appeared simultaneously, depending on the actual conditions of the biotope and the access to nutrients. As far as the evolution of a postglacial landscape was concerned, different species settled in the present day steppes, others inhabited the tundra and a few lived in both the present tundra and steppe habitats. Subsequently they have become differentiated into independent but related steppe and tundra forms (as subspecies or species).

The realism of this picture is confirmed not only by paleontological material. Landscapes analogous to those about which we have been speaking exist at present. In northern Yakutia and the far north-east of the USSR within the arctic circle at the very limit of the tundra and in places within its area, we may encounter associations of the steppe type or what could very well be called a 'tundra–steppe'. They do not occupy large areas but are distributed over fairly restricted surfaces where large steppe animals are unable to survive but where the majority of plants and insects are typical of a steppe. Thus we can find in the plant cover such species as the grasses *Stipa decipiens, Festuca lenensis* and *Agropyron cristatum* and in the soil larvae representatives of the typical steppe genus *Eodorcadion* of the capricorn beetle family.

It is not entirely clear to what extent these associations should be considered relict or contemporary. Their existence is, in any case, supported by the extremely continental climate, which is enforced by the local climate in inter-montane depressions and valleys. In these areas, especially in northern Yakutia, the summers are very hot with temperatures often around 30 °C which increases the drought. It is interesting that small areas of tundra–steppe can be met with as far north as on Ostrov Vrangelya (Yurtsev, 1974).

Some indigenous forest species also display adaptations to tundra conditions. It is obvious that all dendrobionts, i.e., animals closely associated with trees, will be absent from the tundra, but some typical mammalian forest species such as the wolverines (*Gulo luscus*) or the elk (*Alces alces*) will occasionally venture out onto the tundra from the south. The reindeer (*Rangifer tarandus*), however, inhabits the tundra to a far wider extent. At the present time it is preserved in its wild state mainly in the form of a tundra population. Some among them were undoubtedly once indigenous forest animals. This is indicated by the large expanse of forest habitats within its area and its migrations in winter time from the tundra back into the taiga and the forest tundra, and also by the larger dimensions of forest forms and the localities selected for giving birth.

Wild reindeer of the genus *Rangifer* occurred at the beginning of the Quaternary period in the forested areas of North America and was then widely distributed within the boreal landscape. It penetrated into Eurasia most probably over the Beringia connection. The reindeer are not definitely associated with forest vegetation. They feed mainly on grasses, lichens, shrubs and dwarf shrubs. From the point of view of nutrition, the summer tundra is even more favourable for the reindeer than is the taiga.

Basically the particular biotopes play an important role for the animals and plants of the tundra. Thus, soil animals are less dependent on the climate than those forms living above ground. Climatic differences are levelled out and toned down in the soil. Its inhabitants are not subjected to the effects of the maximum and minimum temperature extremes of the air. This is also one of the reasons why soil animals are so widely distributed. For instance, collembolans and bugs in the tundra zone belong generally to polyzonal species. Invertebrates living above ground, on the other hand, have long since developed various specialized adaptations for survival under the severe conditions in the tundra (see Figs. 38, 39).

The habitat conditions in boggy areas within the different zones are almost always stable as a result of which their species of animals and plants have been able to penetrate without difficulties into the tundra mires. As examples may be mentioned the well-known inhabitants of the temperate belt such as marsh marigolds (*Caltha palustris*) and golden saxifrages (*Chrysosplenium alternifolium*), and the yellow wagtail (*Motacilla flava*) as well as the snipe (*Gallinago gallinago*). There are many fine examples of invaders into the tundra zone among the inhabitants of sandy and gravelly beaches of the rivers and lakes.

Southern species penetrate into the subarctic along well-warmed

Fig. 38. A female crane-fly, *Tipula glaucocinerea*, depositing her eggs into the lichen–moss tundra sward.

slopes, especially sandy ones, on which herbaceous associations grow. For instance the heat-loving common thyme (*Thymus serpyllum*) is met with as far north as the typical tundra zone. Thanks to such habitats of the tundra zone some groups of organisms can flourish which are actually thermophilous and sun-loving. The family of hoverflies (the Syrphidae) may serve as examples, these are represented in the subarctic fauna by many brightly coloured species feeding on nectar and pollen.

Thus, animals and plants filling differently formed intrazonal biotopes but of southern provenance and not being specially adapted to living under arctic conditions, are able to cross the climatic boundaries and settle in the tundra landscape. This is an ecological pathway of adaptation to the environment, which in the choice of suitable habitats also includes the avoidance of unfavourable influences.

The physiological mechanism of adaptation is more important. Of course organisms have two means of response to severe conditions: to resist them or to submit to them. The former is, for example, expressed in the warm-bloodedness of birds and mammals, which are able to maintain a constant internal temperature. This requires, however, that they have a constant influx of energy in the form of food. The second response takes the form of cold-bloodedness as in the invertebrates. These are able to stay active only at fairly high temperatures but are still able to survive in an inactive state. The former response may be called an active pathway of

Fig. 39. A characteristic representive of the insects in the arctic tundra, the crane-fly *Tipula carinifrons*, a short-winged female.

adaptation to the environment, the latter a passive form. Elements of either are peculiar to most different groups of organisms and invertebrates as frequently display characteristics of the active type as mammals show a passive mode of adaptation.

The morpho–physiological adaptation of the active type in general implies an increase in vital activities and regulatory processes as well as an acceleration of growth rate and development. Such an adaptation on the whole improves the resistance to unfavourable effects of the environment and at the same time reduces the possibility of existence under extremely adverse conditions which may take the form of a dependence on critical thresholds of decisive environmental factors.

Passive ecological–physiological adaptation subjects the vital activity of an organism to the surrounding environment and at the same time increases its resistance to unfavourable effects and promotes a more economic utilization of energy. For many animals, both warm-blooded and cold-blooded, this is a more acceptable pathway of adaptation to the severe conditions in arctic lands.

The papers by Shvarts and his collaborators have demonstrated that the elements belonging to the active pathway of adaptation, that is, more intense function of the heart, increased fertility, accelerated rates of growth and development, under subarctic conditions are peculiar mainly to widely distributed species, newly arrived from the south. All of these do not, of course, reach as far as the northern limit of the tundra; most settle only in its southern areas. In contrast, typically arctic (and also specialized high alpine) species of mammals and birds usually have slower metabolism, lower fertility and slower rates of growth and development. These facts had been refuted long before any ideas were expressed that it is especially typical for arctic vertebrates to have such traits in their biology as stipulated for greater autonomy of the organism such as a higher metabolism, accelerated development, etc.

Similar characteristics have been discovered also for the invertebrates (Chernov, 1975). It can be stated that the characteristics of most typical high arctic species are slower development, often extended over several years (as many as five), a reduced intensity of feeding and of growth, not too high fertility and a lack of clear-cut seasonal correlation of the life cycles (development may be completed during one season, but under unfavourable conditions it may be extended over several seasons). All these traits promote increased stability of the organism under unfavourable weather conditions and a more economical use of energy, etc. Among the invertebrates in the tundra, the active pathway of coping with the environment, that is raised fertility, accelerated growth rate is, however, also peculiar mainly to southern species or to representatives of generally southern groups. Many species of insects belong here, e.g., leaf cutters, weevils and Diptera.

As a consequence of the slower developmental tempo of many eu-arctic species the annual increase of their populations is small. As a result productivity of tundra associations is also, in general, rather low. At first glance this is contradicted by the rather high level of accumulated biomass. For instance, the number of invertebrates per unit area in the arctic tundra may be two to three times that further south where the conditions are much better (Chernov, 1978). However, in contradiction to this it has to be said, that because of the accumulated biomass of soil inhabitants, prolongation of their life cycles ensues as well as a slower turn over of generations. Concentration of the biomass leads, thus, to a generally lower productivity, that is, to a very low annual increase in populations.

Associated with active and passive pathways of adaptation to arctic conditions may be the peculiar systematic composition of the animal associations. The majority of the large groups of animals prospering within the tundra zone belong phylogenetically to a rather low level and are characterized by a definite primitiveness. Thus, among the hordes of insects thriving best in the arctic landscape we find the collembolans, one of the most primitive forms within this class. Among those coping especially successfully with the tundra, are the old and primitive Diptera, the craneflies (the Tipulidae) and the chironomids, while to the beetles belong the carabids and the staphylinids, and to the Hymenoptera belong the miners.

In the Arctic, within the class of birds, there are the most different forms of species, often also belonging to primitive and ancient groups such as the divers (*Gavia*), the geese (*Anser, Branta*), the snipe and waders, gulls, auks (Alcidae) etc. The representatives of just such groups are able to cope even at the highest latitudes. More progressive groups in the arctic fauna are represented by individual genera or species, often very well adapted to severe conditions, but still not reaching the highest latitudes. For instance among the insects are various advanced Diptera, among the beetles some leaf cutters, among the Hymenoptera bumble-bees and among the higher animals some species of the small passeriform birds.

These features of the components of the fauna of the Arctic are probably caused primarily by the adaptive value of the passive type of adaptation in severe conditions. The primitiveness, and linked with that, the ability to adjust fully to the environment in the worst conditions, are the most advantageous.

7

Distribution of animals and plants

The distribution area of a species results from the effects of climate, landscape and biotopical conditions, interrelationships with other closely related taxa, moisture conditions and competition as well as changes in the environment during the past, etc. In general the area reflects the history of a species, its point of origin, its pathways of dispersal, as well as its present conditions. Historical factors are often revealed by longitudinal boundaries and present factors mainly by the latitudinal ones, that is, the manner in which the species has colonized zones, subzones, etc. It is, of course, almost impossible to separate historical factors from the present. Thus, the Siberian lemming (*Lemmus sibiricus*) is distributed from arctic areas of North America and Chukotka to the shores of Beloye More, but further west, on Kol'skiy Poluostrov and in Scandinavia a closely related species takes over (*Lemmus lemmus*). In this case the boundary between the two species of the same genus may be due both to local factors as well as to pathways of dispersal and behavioural antagonism between them which gradually came to play a role in the past and definitely does so at present. The Siberian lemming is a species of the typical tundra. On Kol'skiy Poluostrov and in Scandinavia only forest tundra and alpine tundra have developed which are less favourable for it, but to which, obviously, the Norwegian lemming is better adapted.

The role of the history of a species, its point of origin and its dispersal routes are especially well demonstrated in such cases where the area extends over a territory covering several zones but at the same time has distinct latitudinal borders. The velvet scoter (*Melanitta fusca*), belonging to the large pochards (*Aythyinae*), is a common species of the European tundra. In Scandinavia, the European parts of the USSR and in western Siberia, this scoter occupies tundra, forest tundra and taiga habitats and sites within the forest–steppe region, but in the east its area is limited by a line running from lower Yenisey to upper Ob (see Fig. 40).

In the Siberian tundra a single species of earthworm, *Eisenia nordenskioldi*, is found from the Yugorskiy Poluostrov to Chukotka. Towards the south it is distributed as far as Primor'ye, Mongolia and even into China. It occurs also in eastern Europe and in the Caucasus. However, it is absent from the central and western parts of Europe and none are found in the tundra of the Kol'skiy Poluostrov and Scandinavia. Obviously, this species dispersed westwards on a broad front from the tundra toward the steppe. It had no competition within the Siberian territory and there it became the single species of its family. But in Europe, where there has always been a rich and variable fauna of earthworms, its area of distribution came to a halt.

A fine example of where the point of origin and the dispersal route of a species is known during the formation of its area is the case of the moss campion (*Silene acaulis*), a representative of the family of pinks, Caryophyllaceae. This small plant with very short peduncles forms a small tuft in the tundra, which will grow compact and build up in the course

Fig. 40. Distribution of the velvet scoter, *Melanitta fusca* (after *Birds of the USSR*, vol. 4).

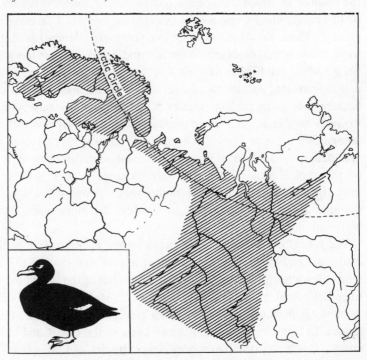

of time a 'cushion' covered with bright pink flowers (see Fig. 41). The moss campion is widely distributed over the tundra and in the high mountains of North America and Europe, but in Asia it is found only in the north-east in Chukotka and Kamchatka as well as on Komandorskiye Ostrova. The area of this species forms a disrupted circle around the Pole with a gap from Yamal to the Kolyma. The botanists consider it an ancient alpine species, originating from the mountains of Europe and dispersed into the tundra and over the Atlantic into North America and from there to north-eastern Asia.

Species distributed in the Arctic and further south in alpine areas are called arctic–alpine (see Fig. 42). Among present-day arctic–alpine species it is possible to distinguish those cases of distribution where the arctic species has been able to penetrate from the tundra environment to the high mountains because of present conditions. Many typical tundra species

Fig. 41. Distribution area of the short peduncled moss campion, *Silene acaulis* (after *Flora Arctica USSR*, vol. 6).

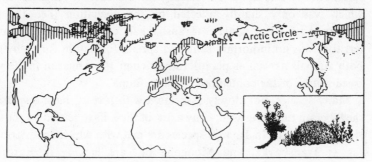

Fig. 42. Typical arctic–alpine distribution of the 'Lloydia', *Lloydia serotina* (Tolmachev, 1974).

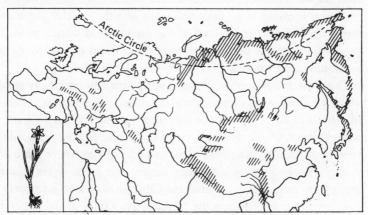

occur in the Siberian mountain ranges, Khrebet Verkhoyanskiy, Khrebet Cherskogo and the Kolymskoye Nagor'ye and on Plato Putorana. The creeping willow (*Salix reptans*), common all over the tundra, is also met with everywhere in the high mountains of northern Siberia. A plover typical of the tundra, i.e., the lesser golden plover (*Charadrius dominicus*), occupies also the Siberian mountain ranges as far as Baykal and Amur. In the mountains directly adjacent to the tundra zone, e.g., on the northern spurs of the Putorana, it is possible to find practically any animal or plant common to the tundra. In contrast to the areas of typical arctic–alpine species, such distributions are due to similarities in the present conditions existing on the tundra and in the alpine environment. The physical closeness of these areas promotes a constant mutual exchange of species.

In the tundra zone of Eurasia there are many species the areas of which reach from Transbaykal, Mongolia and Altay to the arctic coast and often extend west toward the Urals or the Pechora river. Animal species with such areas belong to the so-called Angaran or East Siberian fauna. The formation of this range correlates with the ancient Angaran continent which was only slightly affected by the glaciation (i.e., the present territories of Altay, Sayan, Transbaykal, etc.). In ancient times this territory was an important centre of species formation for many groups, both animals as well as plants. In particular the Angaran fauna was the basic source of the contemporary taiga fauna.

Many species, not closely associated with forests, have dispersed into the Siberian tundra as well. Only a few of these have settled on the western side of the Urals. In Taymyr species are met with which are distributed as far as Transbaykal and Mongolia but are lacking in Europe. For example, the weevil *Phytonomus ornatus*, the larvae of which develop in the buds and flowers of leguminose plants, and species in the high tundra such as the centipede *Monotarsobius alticus* and the sawfly *Pontania crassipes*, forming large galls on leaves of willows, etc. These examples show that animals and plants of the arctic world were formed in close connection with the processes of faunal and floral genesis over enormous areas, including some of the present-day climatic belts.

There are many species in the tundra with comparatively small areas which are oriented in a longitudinal direction, crossing over zonal borders. In other words, at first glance they appear associated with some type of sector but not with a specific type of landscape. Species with this kind of area are especially numerous in north-eastern Asia where there are areas strongly broken up by mountain ranges and large rivers isolating the species. As examples may serve the Chukchi sandwort (*Arenaria tschuktschorum*) distributed from the Chaunskaya Guba to the upper Kolyma and Anadyr rivers; the Dahurian pasqueflower (*Pulsatilla dahurica*) met with from the east Siberian tundra between Lena and Anadyr' rivers down to the Amur river; the short-billed dowitcher (*Limnodromus griseus*), inhabiting an area from Chukotka and the Alaskan tundra down into the taiga zone of North America towards the south. These are examples of typical endemic species, that is, inhabitants of strongly delimited areas. Such areas may represent localities where either the species originated or they may be the only remnants left of a once much larger distribution area.

Examples of endemic species occur where a species exists within an area completely enclosed within the limits of the tundra. Here two phenomena combine, the relationship with the area as such and the association with a particular type of landscape, in this case, the tundra. Within the borders of the tundra zone there are species with very restricted areas. One representative of the Cruciferae, *Draba kjellmanii*, is known only from Vaygach and Novaya Zemlya. Another species of this family is *Cardamine hyperborea*, found only in the tundra along shores between the Chaunskaya Gubu and Chukotka as well as on Ostrov Vrangelya. Ushakov's poppy (*Papaver uschakovi*) occurs only on Ostrov Vrangelya and in Chukotka. The shoveler plover (*Eurynorhynchus pygmaeus*) inhabits the north-eastern shores of the Chukotka Peninsula while the buff-breasted sandpiper (*Tryngites subruficollis*) occurs there as well as in Alaska.

However, in the tundra zone such species are comparatively few. More often the areas include more or less important portions of zones. One of the most characteristic species of the arctic flora is the alpine whitlow grass (*Draba glacialis*) distributed from Poluostrov Kanina to the Novosibirskiye Ostrova but absent from all other tundra zones (see Fig. 43). The

Fig. 43. Distribution area of one of the most typical tundra plants, the crucifer *Draba glacialis* (after *Flora Arctica USSR*, vol. 7).

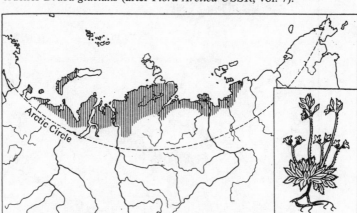

spectacled eider (*Somateria fischeri*) nests on the coast of the Arctic Ocean from the Olenek river all the way to Alaska.

Many species of animals and plants are distributed throughout the tundra zones of Eurasia and America. Such a distribution is called circumpolar (see Figs. 44 and 45). Typical circumpolar areas of distribution are characteristic of the arctic fox (*Alopex lagopus*), the snowy owl (*Nyctea scandiaca*), the snow bunting (*Plectrophenax nivalis*), the king eider (*Somateria spectabilis*), the brent goose (*Branta bernicla*), the long-tailed skua (*Stercorarius longicaudus*), etc. About 40 % of the arctic plant species have circumpolar distribution areas. Species with similar, circumboreal areas are also found in the taiga zone (wolverine, *Gulo gulo*; squirrel, *Sciurus vulgaris*; elk, *Alces alces*) but they are considerably fewer than those in the tundra. Only a few of the typical steppe inhabitants exist on both continents. In the Tropics still fewer genera are present in both hemispheres. One of the reasons for this is the geological history of the present continents. According to the well known theory of A. Wegener they were

Fig. 44. Area of the crucifer *Draba subcapitata*, a circumpolar high-arctic species (Tolmachev 1974).

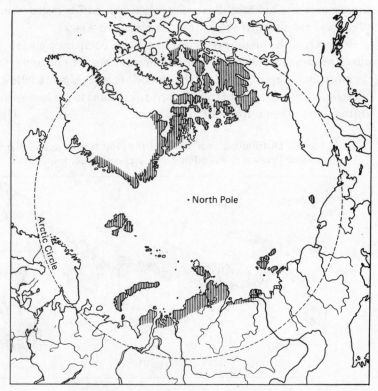

formed as a result of a single ancient continent splitting apart. This started in its southern part with Australia, South America and South Africa being formed. Asia and North America remained united in the area of the Bering Straits until recent geological time. During the Cenozoic they were united and separated repeatedly and an exchange of floras and faunas from both continents could take place over the so-called Beringia area.

Here it was, however, a matter not only of an opportunity for exchange of species. In the tundra zone there is usually a high percentage of species with wide, zonal areas of distribution irrespective of whether they can cross over from one continent to another. The reason for this is still insufficiently understood. The colonization of the tundra zone and the formation of circumpolar areas are no doubt also favoured by the comparatively short distances, the compactness and the closeness of the arctic areas which result from the shape of the Earth. Thus, the length of the 70th parallel is one third that of the Equator. This important fact is often overlooked when considering biogeography.

Fig. 45. Area of the grey plover, *Pluvialis squatarola*, a typical inhabitant of arctic and typical tundra subzones (after *Birds of the USSR*, vol. 3.)

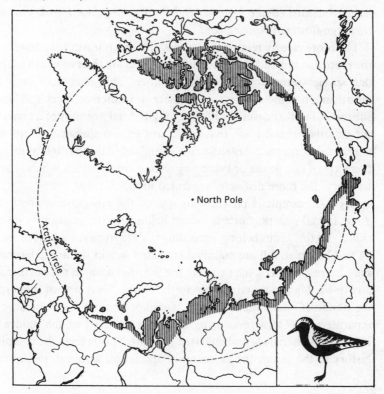

Of great importance for zonal distribution is the poverty of the fauna and flora and the not very high specific variation. This weakens competition between different species and actually favours colonization within the limits of the zones. This may happen both more successfully and faster, the fewer species there are with a similar biology and occupying the same kind of habitat (or ecological niche). Circumpolar distribution is definitely more typical of species not having close relatives and being formed of separate, independent monotypic genera (i.e., genera with a single species). Examples are the arctic fox (*Alopex lagopus*), the snowy owl (*Nyctea scandiaca*), the Lapland bunting (*Calcarius lapponicus*), the snow bunting (*Plectrophenax nivalis*) and the long-tailed duck (*Clangula hyemalis*).

It is evident that well isolated species, being separated by distinct adaptive peculiarities, do offer competition towards other species in the process of colonization of the tundra zone. But the wide zonal distribution of some species may explain that there are, no doubt, many ancient inhabitants of the tundra landscape sharing a long history with it.

Some genera, in general involving only a few species, are, in the tundra zone, represented by a single species only which also is widely distributed and sometimes circumpolar, e.g., Bewick's swan (*Cygnus bewickii*), rough-legged buzzard (*Buteo lagopus*) and the plant species *Pyrola grandiflora*, the large-flowered wintergreen.

The more closely related the species are which inhabit the tundra zone, the more distinct and the better differentiated are their areas. In such cases one species often substitutes for another. This phenomenon is called vicariism. In some of these cases there are different species in different subzones, in other cases in different longitudinal portions of a zone. Often the vicariism is not total, that is the area of one species overlaps that of another, but on the whole the rule is undoubtedly that the more species there are of one genus or belonging to one family which inhabit the same territory, the more distinctly separated are their areas.

A good example is provided by one of the most characteristic of the arctic faunal genera, *Calidris*, which belongs to the sandpipers (Scolopacidae). To this genus belong, according to data from various authors, some 14 to 17 species. All are mainly distributed within the arctic territory: in the tundra zone, to some extent in the forest-tundra, in the northern taiga and in the high mountains close to the Arctic. Only one of these species, the dunlin (*Calidris alpina*), has a circumpolar area. The rest have quite separate areas, some occupying an important part of the tundra zone, others being narrowly endemic (see Fig. 46). This reflects the complicated history of the genus, the processes of speciation within it, the divergence of characteristics, the occupation of different ecological niches and the dispersal in different directions, etc.

The majority of the species within this genus inhabit both sides of the Bering Strait – Alaska as well as north-eastern Asia (see Table 8). Within a considerably smaller area in the eastern subarctic, including Ostrov Vrangelya, Chukotka and the tundra as far west as the Kolyma river, 13–14 species of this genus are met with, in Taymyr nine and in the western Siberian tundras (Yamal, Yugorskiy Poluostrov and Novaya Zemlya) only four species occur. There is no doubt that north-eastern Asia stands out as an area of the most intensive species formation and possibly it may also be the centre of origin of this genus, from where it dispersed eastward over the American continent and westward over Eurasia. Where there are sufficient species of sandpiper, they inhabit different biotopes or different subzones. In Taymyr there are a total of nine species but in each of the subzones there are only from three to six. With regard to the distribution of the species belonging to this genus it is clearly apparent that all the forms are territorially delimited longitudinally, latitudinally and biotopically. Thereby the role of historical factors seems especially distinct, that is the place of origin and the direction of dispersal of the species in question.

Where it is not a question of a single genus but of a group of higher taxonomic rank, a lesser regularity and a greater variation in distribution and in correlation between specific areas can be established. Thus, there are ten species belonging to the subfamily of geese in the Eurasian tundra.

Fig. 46. Distribution areas of four species of sandpipers in Eurasia. 1, The sanderling (*Calidris alba*); 2, the little stint (*C. minuta*); 3, the sharp-tailed sandpiper (*C. acuminata*); 4, Baird's sandpiper (*C. bairdii*).

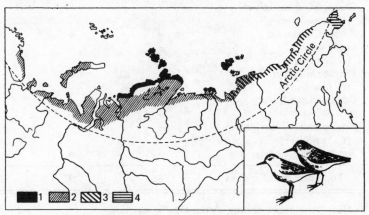

They are distributed in four genera: three species of true geese, i.e., the white-fronted goose (*Anser albifrons*), the bean goose (*A. fabalis*) and the lesser white-fronted goose (*A. erythropus*); four species of brants, i.e., the brent goose (*Branta bernicla*), the barnacle goose (*B. leucopsis*), the Canada goose (*B. canadensis*) and the red-breasted goose (*B. ruficollis*); and two monotypical genera, i.e. the blue or snow goose (*Chen caerulescens*) and the emperor goose (*Philacte canadica*). Their distribution within the tundra zone is very complex. The three species of *Anser* inhabit the entire extent of the Eurasian tundra zone and their areas overlap in general. But each displays specific characteristics in respect of zonal distribution. The

Table 8. Distribution of sandpipers in Eurasia

Species	Novaya Zemlya, Yugorskiy Poluostrov	Taymyr	Chukotka
Curlew sandpiper (*Calidris ferruginea*)	−	+	+
Dunlin (*C. alpina*)	+	+	+
Little stint (*C. minuta*)	+	+	+
Temminck's stint (*C. temminckii*)	+	+	+
Rufous-necked sandpiper (*C. ruficollis*)	−	+	+
Long-toed stint (*C. subminuta*)	−	−	+
Baird's sandpiper (*C. bairdii*)	−	−	+
Pectoral sandpiper (*C. melanotos*)	−	+	+
Sharp-tailed sandpiper (*C. acuminata*)	−	−	+
Knot (*C. canutus*)	−	+	+
Great knot (*C. tenuirostris*)	−	−	+
Purple sandpiper (*C. maritima*)	+	+	−
Beringian sandpiper (*C. ptilocnemis*)	−	−	+
Semi-palmated sandpiper (*C. pusillus*)	−	−	+
Sanderling (*C. alba*)	−	+	−
Total number of species	4	9	13

Note: + : species nesting within area in question
  − : species not nesting there

white-fronted goose is met with from the southern to the northern limit of the tundra. These geese nest not only in the tundra zone but also in the forest–tundra and the taiga territories. The lesser white-fronted goose is distributed mainly in the southern tundra and the forest–tundra. The brent goose is a bird of the high-arctic coast. The other birds have strictly regional areas. The emperor goose nests along the arctic coast of the Atlantic, the Canada goose within the arctic and more temperate latitudes of North America. The red-breasted goose is endemic to Taymyr and Gydanskiy Poluostrov and the barnacle goose to Chukotka and western Alaska. The blue or snow goose is distributed in the coastal tundra from Ostrov Vrangelya all the way to Baffin Island. Within this group, there is much less mutual interrelationship in the distribution of the species.

The vicariism of species with similar biology is especially distinct. Fine examples can be found in the plant genus *Trollius* (the globe-flowers; see Fig. 47) and in two species of birds, the lesser golden plover and the golden plover (*Pluvialis dominica, P. apricaria*), inhabiting the dry tundra plakor.

One of the most interesting biogeographical phenomena within the fauna and flora of the Arctic is the breakdown of distribution areas into disjunct areas or those occurring in separate parts of a territory. These are very typical of the inhabitants of the high latitudes, the polar desert and the arctic tundra. In such cases the disruption of the area can almost always be explained by the fact that the polar desert and the arctic tundra are represented by fragments on islands and peninsulas in the Arctic basin. Such distribution areas are typical of the knot (*Calidris canutus*), the sanderling (*C. alba*) and the purple sandpiper (*C. maritima*). Analogous

Fig. 47. Distribution areas of globe flower (*Trollius*) species in northern Eurasia (after *Flora Arctica USSR*, vol. 6). 1, *Trollius europaeus*; 2, *T. apertus*; 3, *T. asiaticus;* 4, *T. boreosibiricus;* 5, *T. membranostylis.*

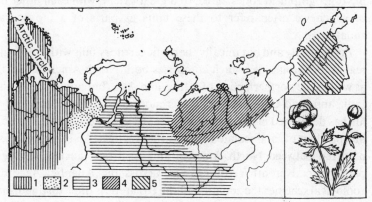

fragmentation of the distribution area occurs in the arctic avens (*Novosieversia glacialis*), a typical high-arctic plant species. In that case the cause of the disruption of the area is the occurrence of broken-up landscape.

As already mentioned more than once, coastal species are characteristic elements in the fauna and the flora of the Arctic. These are various kinds of gulls, auks and guillemots, eiders and brent geese as well as various insects and plants. Some of these are distributed widely within the subarctic and especially the arctic landscapes but there is a 'gap' on the coast along the Kara and Laptev Seas. They are also often lacking in Taymyr. Such gaps in their distribution are, for example, common for the eiders and some species of coastal flies, distributed along watercourses flowing to the seashore. There is no basis for the statement that this is caused by historical factors or by irregularities in dispersal. In these areas the coastline of the continent reaches very high latitudes, in Taymyr as far as the 77th parallel. The sea surrounding it is very cold and frozen. The littoral is practically lifeless. There are no brown algae, so common in other portions of the arctic coast, and in the shallow waters no mass colonization by mussels or other invertebrates. In other words, there is no food in those coastal waters for eiders that feed on invertebrates or for flies and their larvae which develop on decomposing algae.

One result of the studies of distribution areas (their dimensions, interrelationship and history) is the zoogeographical and phytogeographical zonation of the Earth. The basic problem is to reflect the developmental processes within the plant and animal environment, that is, to distinguish the territories where species formation occurred more or less in isolation. The largest category of zonation is the region e.g., the Australian region, the neotropical region (Central and South American), and the Ethiopian region (Africa south of the Tropic of Cancer), etc. Zoogeographical and phytogeographical zones agree, to a great extent, with each other because of which we often refer to these units as parts of a biogeographical zonation.

Zoologically and botanically the arctic territory falls within a single large region, the Holarctic, which comprises both North America and Europe as well as the northern parts of Africa and Asia (but not its southern parts, India and Indo-China). Over all this enormous territory plant and animal life has developed in the course of extended geological periods more or less independent of, and isolated from, other regions. An exchange of species occurred between North America and Asia over ancient Beringia and partly also over other routes, e.g., across the Atlantic. According to zoological schemes the zonation of the Holarctic is often distinguished into two separate subregions, the Palaearctic (Eurasia) and the Nearctic (North America). But in our case we may retain the concept of a single Holarctic. It is sometimes also called a 'kingdom' and subdivided into 'areas', 'subareas' and 'provinces'.

The majority of zoologists and botanists include all the arctic territories within the single arctic region of the Holarctic. There is good reason for this. The majority of species and genera of plants and animals, distributed within the territory of the tundra zone in both hemispheres seem, of course, to emphasize the floristic and faunistic unity of the arctic region. However, when examined more carefully, the problem is not so simple. Examples have been given above where species inhabit relatively small portions of the tundra but larger parts of more southerly zones. Such species are especially plentiful in north-eastern Asia. Their areas often encompass the north-western extremes of America as well. These territories were the centres of formation and dispersal of species belonging to a very large number of animal and plant groups. In this connection an independent biogeographical subregion, Beringia, has been distinguished. It covers north-eastern Asia, Alaska and adjacent areas, including both the tundra and the taiga south of it (Yurtsev, 1974).

Many species of plants and animals reach their western limit on Taymyr at the lower reaches of the Yenisey river. This is one of the most important biogeographical boundaries. Some such species are typical of the tundra, others inhabit the taiga further south as well as the Siberian mountain areas. Some examples were given above of such distributions. In this connection an independent eastern Siberian (Angaran) subregion has been distinguished as a part of the Palaearctic, uniting within it both tundra and taiga areas. The well known zoogeographer I. I. Pusanov promoted this opinion.

The distinction of such categories as the Beringian and the Angaran subregions complicates the idea of floristic and faunistic unity of all the arctic territories. At present, criteria for establishing biogeographical categories have not been sufficiently well worked out. For example, it is not clear how large the percentages of endemic species, of typical species or of other taxonomic units must be in order to bestow the rank of region, subregion or province on this or that territory. Thus, in contrast with other botanists, Takhtadzhyan (1978) considers the entire arctic area as nothing but a province of the Holarctic. However, such contradictory ideas are not built on actual data regarding the distribution of plants and animals but on basically faulty concepts regarding zonation. It is, in particular, still not clear how regions should be correlated when distinguished on the basis of the structure of the vegetation cover, the associations in total and the genesis of the fauna and flora. Investigations at this level regarding the tundra zone will no doubt be very valuable in working out the general principles of biogeographical zonation.

Species with very different areas and zonal delimitations can all be called 'arctic'. Thus the Siberian lemming (*Lemmus sibiricus*) inhabits all the tundra zone and a large portion of the polar desert, while Middendorff's vole (*Microtus middendorffii*) lives mainly in the southern tundra and the forest–tundra. The small snowgrass, *Phippsia algida*, is always widely distributed within the arctic area, but occurs locally outside its main area extending into habitats within more southerly, alpine areas. It plays an important role in the vegetation covering the polar desert and is abundant in the arctic tundra subzone. On the typical tundra it grows in the coldest habitats, e.g., around snow drifts. This species is a typical inhabitant of the high Arctic. Another grass, the alpine foxtail (*Alopecurus alpinus*), grows in the southern areas of the tundra zone on north-facing slopes, in habitats percolated by cold water and in depressions with long-lasting snow. In the arctic tundra subzone it is abundant almost everywhere and forms a continuous turf, looking almost as if it had been sown there. The little stint, *Calidris minuta*, is a bird with a characteristic massive occurrence within the typical tundra, but it is lacking or rare in the arctic and southern parts of the tundra.

Thus, many plants and animals are correlated with different latitudinal belts of the tundra, that is, subzones, or even with parts of them. Consequently, the general concept of an 'arctic species' must be better distinguished and defined in detail. It is possible to make such a distinction by using in the main four variants of zonal distribution, while applying prefixes from the Greek language: hyper-(high), eu-(typical), hemi-(half), and hyp(o)-(below), i.e., hyperarctic, euarctic, hemiarctic and hyparctic. The first category includes many typical inhabitants of the polar deserts, the other three reflect relationship to the tundra subzones. These categories of zonal distribution appear more clear-cut with respect to the mammals, the birds and the higher insects, but less so for the flowering plants. The terms apply only to individual species of lower plants and they do not seem suitable at all for a number of groups belonging to the microscopic invertebrates.

'Hyperarctic' or high-arctic species extend mainly over the islands in the Arctic Basin within the zone of the polar desert and in the subzone of the arctic tundra. The high arctic vertebrate animals are, in the majority of cases, closely associated with the sea: polar bears, birds commonly found on bird cliffs, such as the little auk (*Alle alle*) and the ivory gull (*Pagophila eburnea*). In this group may be included also the Brent goose

(*Branta bernicla*) and some sandpipers – the knot (*Calidris canutus*) and the purple sandpiper (*C. maritima*) – although locally their areas extend into much lower latitudes. Among plants we could mention some species which are clearly associated with the highest latitudes although the most typical representatives of the hyperarctic area are not so restricted. An analogous example is the ivory gull. Examples of plant species best adapted to the polar desert may be grasses such as *Phippsia algida* and the small crucifer *Draba capitata* (see Fig. 44).

'Euarctic' or truly arctic species inhabit the northern belt of the tundra zone, that is, the arctic subzone and part of the typical subzone. Some of these species are distributed over all the tundra subzones down to the southern limit but their ecological optimum lies clearly within the arctic and the northern part of the typical tundra. Typical representatives are the snow bunting (*Plectrophenax nivalis*), the curlew-sandpiper (*Calidris ferruginea*), the snowy owl (*Nyctea scandiaca*), the collared and the Siberian lemmings (*Dicrostonyx torquatus*, *Lemmus sibiricus*) and the cranefly (*Tipula carinifrons*; see Fig. 39), and among plants Sabine's buttercup (*Ranunculus sabinei*), alpine whitlowgrass (*Draba alpina*), Edward's starwort (*Stellaria edwardsii*) and some species of saxifrages, e.g., *Saxifraga cespitosa*. These species display the most characteristic traits of the arctic flora and fauna. Usually they show distinct adaptations to life under severe conditions as, for example, reduced body size, gradual development over several years, adaptations to bare and gravelly soils, exposed habitats, etc.

'Hemiarctic' species, or those of the typical tundra, are characteristic inhabitants of the typical tundra subzone. They do not usually reach the northern limit of the tundra zone and they are rare at its southern border although they often extend further south in the high mountains. To this group belong many widely distributed inhabitants of the tundra, e.g., the dunlin (*Calidris alpina*), the little stint (*Calidris minuta*), the bumble-bee (*Bombus lapponicus*), and plants such as the creeping willow (*Salix reptans*), the mountain aven (*Dryas punctata*) and the arctic bell-heather (*Cassiope tetragona*).

'Hyparctic' species are inhabitants of the southern tundra subzone, the forest–tundra and parts of the northern taiga landscapes. Their areas often extend over narrow belts within the southern subarctic, sometimes far south into the taiga zone and frequently into mountainous areas. Typical representatives are Middendorff's vole (*Microtus middendorffii*), the willow grouse (*Lagopus l. lagopus*), among shore-birds such as the bar-tailed godwit (*Limosa lapponica*) and spotted redshank (*Tringa erythropus*), the lesser white-fronted goose (*Anser erythropus*), the little bunting (*Emberiza pusilla*), and the largest of the tundra spiders, the arctic tarantula (*Lycosa hirta*), as well as the largest of the tundra beetles (*Carabus odoratus*); in addition there are plants such as dwarf birch (*Betula nana*), crowberries (*Empetrum nigrum*) and cloudberries (*Rubus chamaemorus*). In the forested belt there are usually intrazonal biotopes (in bogs and on ridges, crests and hills). North of the southern tundra subzone the number of hyparctic species falls off sharply and in the middle portion of the typical tundra subzone they are very few, while in the arctic subzone hardly any at all.

Not every species can be referred to any particular group. Thus, the majority of lower plants, especially microscopic species and many species of flowering plants, soil animals, insects and even some birds and mammals do not have any noticeable zonal affiliation. Often such species are distributed across many zones, for example, over the entire territory of the Holarctic or of the Palaearctic, and sometimes they are found almost everywhere in the world. Such species are called transpalaearctic, trans-holarctic or cosmopolitan, respectively, or if in association with zones, polyzonal or – rarely – multizonal. Typical examples of polyzonal species occurring within the tundra zone are the stoat (*Mustela erminea*), the seven-spotted ladybird (*Coccinella septempunctata*), and plants such as the golden saxifrage (*Chrysosplenium alternifolium*) and marsh marigolds (*Caltha palustris*). Typical cosmopolitans are also found such as the peregrine falcon (*Falco peregrinus*) and various small soil arthropods, e.g., the armoured mite *Liochthonius sellnickii*, and the springtailed *Cerato-physella armata*, etc.

Areas covering several zones may indicate indifference to changes in the conditions typical of species with ecologically wide ranges. This applies particularly to such species as the stoats and the seven-spotted ladybirds. They are distributed over enormous areas, almost the entire northern hemisphere, and inhabit the most variable kinds of biotopes, deserts, steppes, forests, tundras, alpine meadows, etc. Although they seem to prefer some biotopes (e.g., the stoat frequently settles along the banks of watercourses) more than others, they are able to survive in many different kinds of environment.

Many polyzonal species select, however, very definite types of biotopes. Here belong different shore species colonizing river banks within the tundra and other zones, lakes, edges of bogs and mires, gravelly spits, sand dunes, etc. Plant species such as the marsh marigold (*Caltha palustris*) and some golden saxifrages (*Chrysosplenium alternifolium*) are found almost everywhere within the territory of the USSR along brook banks, in bogs, in wet meadows, etc. The mud-beetle, *Elaphrus riparius*, lives on sandy and gravelly spits along rivers from the forest belt all the way deep into the tundra. Some species of flies have very large longitudinally extended areas, their larvae developing in rotting algae along the sea coasts. Such a fly, *Fucomyia frigida*, is distributed from tropical latitudes all the way to northern Greenland, Novaya Zemlya and Chukotka. The large shore grass, *Leymus mollis*, is a typical inhabitant of coastal sand dunes along which it extends from the Tropics to the Chukotka.

The wheatear (*Oenanthe oenanthe*), a bird belonging to the thrush family Turdidae, is distributed from Africa to the arctic tundra subzone, but everywhere within this enormous territory it nests only in special zonal associations. It builds its nests at the foot of precipices, among boulders, under rocks, in holes, along river banks, in rubble, in villages, etc. In the southern half of the tundra zone, the yellow wagtail (*Motacilla flava*) is common. It is distributed over the entire territory of Asia and everywhere associated with bogs, boggy meadows and bushy thickets.

Thus, within the extension of their enormous ranges, these species are often associated with specific intrazonal habitats within the landscape. Outside their biotopes they are not very numerous, sometimes absent over large areas, but then they may appear again. In general they are widely distributed and independent of zonal borders.

The conditions on coastal sand dunes, in bogs, in wetlands cut off from the water along the coastal belt or on rocky substrates are highly specific and require special and far-reaching adaptations. Although the species may be adapted to such conditions, other factors are of equal importance here. In addition, in such biotopes, climatic variation seems to be toned down. For instance, independently of where river banks occur, in a tundra or in a desert environment, there is always a considerable amount of moisture; at the same time proximity to water has a cooling effect during strong heat and during cold periods a warming influence, since the temperature of water bodies changes at a slower rate than that of air. As a consequence, the similarity of plant and animal habitats there is quite striking and biotopically correlated.

This does not mean, however, that all inhabitants of such biotopes are distributed independently of zonal borders. It is just that they are less dependent on the oscillation of climatic factors than plants and animals of plakor habitats and in zonal associations. The majority of the characteristic inhabitants of wetlands, coastal areas and rocky substrates occur only within a single zone. Often among them, we encounter taxa associated with intrazonal biotopes or endemics distributed over small areas within a single zone. Each such case must, however, be explained on the basis of some particular history of the species in question, its biology, its interrelationship with other species, etc.

In the distribution of plants and animals a major role is played by the extrazonal associations, such as are found beyond the 'proper' borders of zones or subzones, for example, within forest islands on the southern tundra subzone, bush associations in the typical tundra subzone, etc. In the southern tundra subzone many inhabitants of the forest tundra, taiga or other forested landscapes occur. They are, indeed, associated with thickets of bushes or shrubs, or isolated within habitats of forests on the slopes of hills, or in the valleys of rivers or brooks. Two species of globe flower (*Trollius europaeus, T. asiaticus*), the European and the Asiatic, the little bunting (*Emberiza pusilla*), the arctic warbler (*Phylloscopus borealis*) and the willow warbler (*Ph. trochilus*) belong here. Locally, thickets of tall shrubs and even forests extend very far northwards. Thus, in the Bol'shezemel'skaya Tundra islands of willows are met with and in the central parts of Taymyr, close to Ozero Taymyr, willow copses are found, as tall as a man. In such biotopes live many southern animals and plants, not otherwise common in the zonal tundra associations.

The phenomena described here explain the different routes enriching the poor and limited life in the tundra with more southern taxa. No doubt those species, which are widely distributed not only in the tundra but also in more southerly zones, are indeed able to become secondary colonizers of the young subarctic landscapes.

8

Interrelationships between organisms

The infinitely variable interrelationships between organisms are usually referred to some basic types: competition, symbiosis, parasitism, etc. No doubt an important role is played by procedures such as food gathering and by processes by which organisms acquire the necessary energy to stay alive. Green plants (autotrophs) synthesize the bulk of organic matter. In the tundra these are algae (with which the lichens may be loosely included as well), mosses and flowering plants. Organic matter, synthesized in the form of plant material, is called *primary production*. Usually this is calculated in units of weight per year. Production is accomplished both by these plants and by various heterotrophic organisms which are responsible for *secondary production*. The heterotrophs are separable into saprophytes (fungi, some bacteria), which utilize decomposition products of complex organic matter, and holozoics, that is animals using a combination of organic matter as nutrients. The holozoics are in turn distinguished into saprophages (consumers of dead plant material), phytophages or herbivores (consumers of plants), predators (those catching and devouring other animals) and necrophages (those eating carcasses of other animals). All together this concerns problems of production and energy, ecological principles which are as yet very little studied as part of research into tundra associations.

Animals and microscopic plants

Animals feeding on microscopic plants are called microphytophages or microbophages. Their relationship with their food sources, especially under arctic conditions, is very little understood. It is very difficult but important for ecologists to distinguish between consumers of autotrophic microscopic plants (algae) and those feeding on heterotrophic organisms (bacteria, fungi and protozoa).

119

The microscopic algae occupy an obvious position within the tundra associations. As a consequence of the constantly high amount of moisture in the soils, they can always find favourable conditions here. Being characteristic of primary and primitive soils, the algae develop in enormous numbers as the ground warms in the spotted and arctic tundras and also in the arctic desert. In these biotopes they constitute the basic part of the primary production. Due to their abundance, the soils are often coloured green over several square centimetres. In such patches great numbers of various animals exist, which feed on the algae (algophages), such as nematodes, enchytraeid worms, springtails and chironomid larvae.

The larvae of the chironomids are a well-known food of fishes and much appreciated by people keeping aquaria. The majority of the chironomid larvae live in water and are coloured bright red. However, there are also land species which are not as colourful but rather greenish or pale grey. These are especially numerous in the arctic soils (up to several hundreds per square metre), where they feed on both microscopic algae and on aquatic species.

The soil nematodes have extremely variable life forms. Besides typical microphytophages, feeding on algae and even on bacteria, there are also predators and parasites on animals and plants, consumers of fungal mycelia (mycophages) as well as some multivorous (or polytrophic) species. Specialists are able to distinguish the type of food used by different species of nematodes on the basis of the structure of their mouth organs, the shape of their bodies and the contents of their intestines. Microphytophages predominate in the tundra. About half, sometimes as much as 70% or more, of the nematodes belong to them. But there are also many polyphages. Only in the meadow-like tundra associations are the microphytophages comparatively few. There, parasites on higher plants predominate.

The number of nematodes in the tundra ranges from 1 to 5×10^6 specimens per square metre, while in the subarctic meadow-like nival associations it may be as high as 10×10^6. The possible number of consumers of micro-organisms such as algae and bacteria in the tundra and meadow-like communities amounts to about $2 \times 10^6/m^2$, i.e., 1.5 g fresh or 0.5 g dry matter. It is possible to calculate roughly the amount of food consumed by the nematodes feeding on microscopic plants. For this it is necessary not only to know the quantity of these organisms at a given moment but also their productivity, that is the excess quantity produced per unit of time. According to Parinkina (1973) the production of bacteria in the spotted tundra may amount to 2 kg/m²/year in fresh weight. For nematodes, the annual amount of production is of the order of 45 g/m². It is possible that while consuming bacteria, the nematodes assimilate 70% of the edible material but that 90% of the assimilated nutrients are spent on sustaining life and only 10% used for reproduction. Consequently the nematodes consume roughly 32% of the annual production of bacterial mass. There are almost no data on the supply and propagation of algal mass within the tundra. Neither is it known to what extent the nematodes are being killed off. In cases where the nematodes feed mainly on algae, they most likely consume not less than half the production of these microscopic plants.

In addition to the nematodes, somewhat higher animals also feed on algae, in particular those annelid worms belonging to the Enchytraeidae, which measure 0.5–3 cm in length. Generally they consume algal cells, together with detritus and the remains of plants. The aquatic enchytraeids living in wetlands are even more closely associated with the algae. Their intestines are often filled with green and blue-green algae. Armoured mites, springtails and different kinds of larvae of flies and mosquitos also feed on algae. We have already mentioned the chironomid larvae above.

There is still another group of microphytophages which are at present not well known within the tundra zone, that is the protozoas. All that is known is that they can occur in great quantities in tundra soils. They can be observed as microbial accumulations on specially prepared microbiological glass slides. It is, however, not known what kind of life they lead or what they feed on or who feeds on them.

It could be suggested that the algal complexes and their consumers play an important role for life in the tundra zone. Animals which feed on algae, are in turn being consumed by predatory insects, spiders, beetles and even by birds, especially snipe, sandpipers and plovers. The latter are then taken by predatory birds or mammals such as polar foxes or stoats. Thus, a food chain is formed which starts with the microscopic autotrophic organisms. It is particularly important that we learn more about the most northerly variants of the arctic associations, those in the subzones of the arctic tundra and the polar desert. Here the microphytophages form the basis for the production of living matter.

Animals and lichens

The interrelationships between animals and lichens is an interesting but little studied ecological problem. It is especially important with respect to investigation of tundra ecology, where lichens make up the greatest part of the plant cover. They are represented by a multitude of species and by life forms such as those having fruticose bodies with flat or tubuliform thalli (*Cetraria* and *Cladonia*, respectively), piliform bundle-like thalli (*Alectoria*), digitate or so-called vermicular thalli (*Dactylina, Thamnolia*) or foliose thalli (*Peltigera*). There are also those forming mould-like crusts, etc. They are met with scattered among mosses, forming bundles and cushions, or they colonize bare ground and stony substrates.

The interrelationship between animals and lichens in the tundra is always variable. For many of the small invertebrates (e.g., mites or springtails) the lichenous swards and cushions provide satisfactory habitats. Under lichen thalli, where the soil is warmer, or on the sloping sides of rocks and stones, many soil animals are concentrated and assist in the formation of primary soil films. Dead particles of lichens obviously provide food for saprophagous animals. Many birds, e.g., snipe, sandpipers and plovers as well as ptarmigan, use lichens as nesting material. In particular they use the soft white *Thamnolia vermicularis* and the digitate thalli of *Dactylina arctica*. During winter lemmings sometime build their nests of lichens. In the stone-paved arctic tundra and especially in the polar desert where herbaceous plants are widely scattered and often not adequate for nesting purposes, wild animals utilize entire cushions of *Cetraria delisei* for this purpose. Cushions 15–30 cm in diameter can be found under the snow. The lemmings crumble them to construct spherical nests of the remains.

Lichens are very nourishing. Fungal hyphae, forming the thallus, consist mainly of special polysaccharides called lichen starch. There are also other polysaccharides as well as disaccharides. Nitrogenous matter is built up into many amino acids. Also they contain many vitamins, e.g., ascorbic acid, cyanocobalamine (vitamin B_{12}) and nicotinic acid, etc. The mineral matter in the lichens varies considerably. Often the composition and the quantity of the minerals depends on the substrate on which the lichens grow. There are also many so-called lichenous substances capable of different forms of biological action; some are, for instance, able to suppress the growth of some other fungi and mosses or the germination of some flowering plants. They seem to have an effect even on animals but this problem has hardly been touched upon as yet.

The food relationship between animals and lichens is little known. Very little is written about those species of invertebrates for which lichens may be a basic food. For instance, butterflies are known for their endlessly variable trophic interactions with plants but the larvae of only a few species have been mentioned as developing among lichens. Almost nothing is known about invertebrates on the tundra feeding on this group of plants.

Apparently none of the herbivorous birds living on the tundra, geese or ptarmigan, feed on lichens, although small parts thereof may accidentally arrive in their stomachs as a mixture with the roots normally consumed. This is hardly ever mentioned in the literature.

Some species of wild animals are known for whom lichens play a major role as a food during winter. Inhabitants of the east Siberian taiga such as the musk deer (*Moschus moschiferus*) and some other deer consume epiphytic lichens growing on conifers. The wild and domesticated reindeer (*Rangifer tarandus*) is the only species of mammal which normally can subsist for a long period of time while feeding exclusively on this group of plants. They consume different species of lichens but especially the 'Reindeer moss' (*Cladonia*). It has to be stated, however, that the reindeer are specialized for feeding on lichens. They are able to live and propagate while consuming nothing but lichens, but also they manage easily without them. They eat many kinds of herbaceous food with great appetite.

The Reindeer moss (*Cladonia*) is their basic winter food in the arctic, alpine and forest tundras as well as in the northern taiga zone. There, these lichens form closed communities over very large areas. The *Cladonia* is known to be very abundant on Kol'skiy Poluostrov and in Evenkii National'nyy Okrug south of Taymyr where the reindeer stay during the winter (see Figs. 48, 49). There the reindeer dig up the *Cladonia* from below the snow, using their strong cloven hoofs. Thanks to this adaptation the reindeer are able to manage in the arctic landscape. During winter lichens do not die off and do not wilt or lose their nourishing qualities. In a moist environment their texture remains soft and juicy and they are easy to nibble and chew. They also form a continuous cover which allows the animals to get their fill over a small area and this makes them an especially important subnival winter food.

Lichens propagate very slowly; in one year *Cladonia alpestris* has an

Fig. 48. Grazing reindeer (*Rangifer tarandus*) in Taymyr (photo.: A. V. Krechmar).

increment of only 4.8 mm. The productivity of this species is low, about 350 kg ha^{-1} y^{-1}. Semyenov-Tyan-Shanskiy (1977) states that while nibbling *Cladonia* the reindeer use up as large an amount of lichens as will take ten years to regenerate. Therefore the localities of intense grazing by either wild or domestic reindeer will be depleted in the course of 3–4 years or at least start to become sparse or degenerate. In their place other variants of the plant cover, for example grasses, will develop. Only after a period of 6 years freedom from grazing can the Reindeer moss begin to re-cover the

Fig. 49. The annual life cycle of the Taymyr herd of reindeer (*Rangifer tarandus*). 1, Calving grounds and concentration during summer; 2, basic wintering area; 3, autumn migration route southwards; 4, departure from wintering grounds; 5, limit of the tundra.

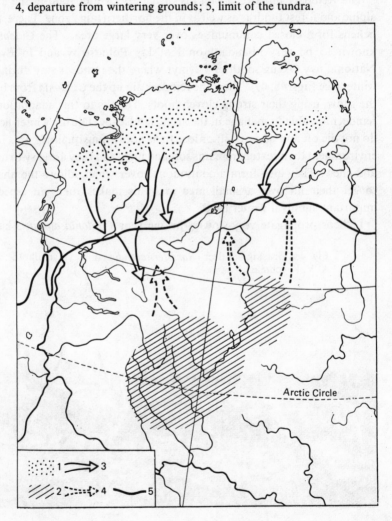

damaged pasture land. For complete restoration of a strongly depleted winter pasture, 15–20 years are necessary. During the summer, fires often run through the *Cladonia*-covered areas. Following such an accident, the restoration may require up to 70 years.

Some peculiarities of lifestyle and behaviour of reindeer must be considered when trying to avert rapid destruction of winter pasture. During the summer, wild reindeer abandon winter pastures and migrate to other sites. Thus, the Taymyr herd wanders to northern Taymyr to calve. When domesticated in the southern parts of the subarctic as, for example, on Kol'skiy Poluostrov, reindeer can be distributed over an immense area.

Lichens are very valuable to the reindeer but are by no means exclusively their favourite food. During summer not only do reindeer feed readily on herbs, especially legumes, grasses, sedges and cotton-grasses but also on the leaves of bushes and dwarf shrubs. They also consume mushrooms and even various small animals such as lemmings and baby birds. The role of lichens in the diet of the reindeer diminishes from the southern areas of the subarctic towards the north. In the most northerly parts of the area, that is in the subzones of the typical and the arctic tundras, they are not very important. There the reindeer feed mainly on gramineous plants and dwarf shrubs.

During summer in the tundra lichens constitute merely a part of the reindeer diet. There are several reasons for this. First of all, in the high latitude tundras there is barely any Reindeer moss. There, other kinds of lichens predominate, mostly those with rough and stiff thalli which are less palatable to reindeer, such as species of *Cetraria* and *Stereocaulon* as well as other crustaceous species, etc. Secondly, lichens do not form a closed vegetation cover in these tundras, such as is especially important for reindeer in winter pastures. Either the lichens are scattered in the moss cover or they grow in the form of isolated patches. During summer these lichens are strongly desiccated, becoming fragile, stiff and prickly. Therefore reindeer consume them mainly on humid days when the thalli swell up with moisture and become softer. The same applies in the forest tundra and the taiga.

Thus the reindeer has adapted to feeding on lichens mainly during the winter. Thanks to lichen forage alone, the reindeer are able to winter over in the forest tundra and the northern taiga. They must, necessarily, leave the typical and the southern tundras during winter when there is much compact snow and where lichens do not form a close cover.

Whether lichens are a food for lemmings is not quite clear. Just like the reindeer they feed readily on mushrooms. It would therefore be assumed that lemmings consume lichens as well, since their thalli are to a great extent formed by fungal mycelia. But there is hardly any information on what role lichens play in their diet. On the basis of long-term research in Yamal and of experimental feeding using various types of diets, Dunayeva (1948) is of the opinion that neither of the two species of lemmings feed on lichens. Different opinions have been presented by Yudin, Krivosheyev and Belyayev (1976) regarding lemmings in the tundras of the Far East. On Ostrov Vrangelya lichens seem to constitute up to 60% of all the fodder stored by the collared lemming (*Dicrostonyx torquatus*). These authors have not, however, proved that the lemmings actually consume all the stored material.

Thus in the tundra lichens, just as in other zones, are not much utilized by grazing animals. The reasons why are not clear. The very close relationship between water content of the lichens and the humidity of the surrounding environment, and the rapid desiccation in a dry atmosphere when they are brittle and very stiff is unfavourable for the herbivores. It is also possible that the specific chemistry of the lichens plays a decisive role, particularly the high concentrations of biological substances contained in them.

The relationship between lichens and insects has to a great extent been predetermined by evolutionary factors. The evolution of the insects is closely connected with that of the higher plants, especially with the angiosperms. Lichens were left outside the basic direction with respect to food specialization of phytophagous insects. Thus, the limited group of insects which has a food relationship with lichens is absent from the tundra.

Animals and mosses

Mosses in the tundra play an important formative role. Above, the reader has been reminded repeatedly of the importance of a moss cover as a factor in the temperature regime of the soil, as a habitat for invertebrates and for the concentration of seedlings and spores for renewing the flora. However, so far very little reliable information exists on the role of mosses as a food source for animals, but there is no doubt that they are consumed.

In the territory of the northern subzone the author has recorded a natural form of trophic interrelationship between insects and the mossy vegetation. This concerns the larvae of *Omphalophora grandis*, belonging to the family Rhagionidae. They feed on the thalli of a species of the liverwort *Marchantia*. These larvae are bright green just like the colour of the thallus. They were found in some localities in northern Taymyr. However, insects do not, as a matter of fact, feed on any true mosses. Among the infinitely variable food relationships between insects and plants the most widely differing tissues are involved, even poisonous ones, but apparently not those of true mosses. It is difficult to believe that there should be any chemical reason for such a situation.

However, as an inevitable admixture, mosses are consumed by larvae and earthworms in thick swards or in soils. At present, very little authentic information exists to confirm that the invertebrates feed on this group of plants. The problem requires specialized research under tundra conditions. Detritus is, in any case, a rather important component in the vegetation cover of the tundra and constitutes one of the ecological riddles.

Tundra birds which consume vegetable matter, that is geese and ptarmigan, never eat mosses; they actually avoid them. When pulling out the juicy parts of species of cotton-grass, sedges and grasses, the geese almost always succeed in not swallowing any moss. Mosses have, however, been found as an unimportant admixture in the crops of willow grouse and ptarmigan.

In the literature we constantly find the information that reindeer consume mosses. Some authors even state that mosses constitute one of the basic components of their diet. This opinion is based mainly on the fact that in the stomach of dead reindeer it is always possible to find remnants of moss. But it is not clear what is the nutrient value to the reindeer. Neither is it clear whether the reindeer consume mosses as such or whether they are an admixture ending up in their stomachs together with fodder plants. Various grasses and bushes have their juicier and more nourishing parts buried in the mosses or covered by the sward and therefore the reindeer inadvertently at the same time swallow some moss when feeding on the other plants. Semyenov-Tyan-Shanskiy (1977), who for a long time has studied reindeer on Kol'skiy Poluostrov, expressed the following statement regarding this problem: 'Two cases are known when moss was actually consumed.... However, some moss is always swallowed together with other fodder.'

In the literature it is mentioned repeatedly that lemmings consume mosses. The idea prevails that mosses are necessary for their normal existence. Such an opinion was held, e.g., by Shvarts (1963). However, Dunayeva (1948), investigating the ecology of lemmings in Yamal, put forward an opposite hypothesis: 'Lichens and mosses are definitely not used by the lemmings as food but only for the construction of their nests.'

As evidence of mosses being consumed by lemmings, the so-called 'bald patches' are often mentioned, that is sites where these animals have 'harvested' the entire moss cover. Such an argument has been presented by, for example, Yudin, Krivosheyev and Belyayev (1976). 'Bald spots' are of course common in the tundra, but they may also bear witness to the contrary. If the winter stores of the lemmings are carefully studied on site, it is possible to note that the bulk of the shredded mosses are left unused. It seems definite that such large quantities of mosses could be used as food by lemmings. In the 'bald spots' the mosses are completely destroyed, literally 'down to the roots', over large areas. If the lemmings were feeding on these mosses, they would tear away their green, living tops and not bother with the basal parts or the dead lower portions. Among remnants in lemming burrows, it is easy to find great quantities of moss with intact tops. For what other purpose then would lemmings harvest such enormous quantities of moss? In the thick moss carpet, covered by deep snow, roots are hidden which are favoured by lemmings – fleshy roots of sedges (*Carex*), cotton-grass (*Eriophorum*), young shoots, leaves and buds of the new growth of dwarf shrubs, herbs and grasses. In order to reach these under snow during winter, the lemmings must literally gnaw their way through the moss sward.

Much work has been spent on the fact that wild animals have an enormous impact on the vegetation of the tundra, for example, turning the moss cover over wide areas 'into dust'. In this way they actually assist in the dynamics of the vegetation and its renewal and raise its productivity. It is well known that moss cover plays a negative role in the propagation of the vegetation (just like the 'tomentum' covering a steppe): it delays the warming of the soil, smothers seedlings, and so on.

Reviewing all the known facts, it is possible to arrive at the conclusion that, typically, lemmings in the tundra do not specifically consume mosses, but swallow them accidentally just as part of the mixture. It may be true that they have some function as 'roughage' in the stomach but this problem needs more detailed research. A great deal regarding the importance of mosses to the lemmings could be cleared up by studies of their particular feeding habits at various stages during their life cycle and over many years. This applies also to a number of other tundra rodents.

Birds, mammals and vascular plants

The mutual relationship between the higher animals and flowering plants in the tundra zone is, just as everywhere else, considerably more variable than their dependence on the spore-bearing plants. Here it is possible to find various forms of food specialization concerned with feeding on special plant parts. This has already been demonstrated clearly for insects. It can also be stated regarding many different birds and mammals of the tundra, that there are herbivorous species which nearly always consume entire plants but during some periods use only special parts. There are, however, also examples of distinct specialization among them, e.g., some granivorous geese and Passeriformes eat only a part of green plants.

The average number of herbivorous vertebrates in the Eurasian tundra zone is relatively high, about 25 species. The most important among them, which are especially widely distributed and most numerous, are the reindeer (*Rangifer tarandus*), the Siberian and collared lemmings (*Lemmus sibiricus, Dicrostonyx torquatus*), the bean goose (*Anser fabalis*) (see Fig. 50), the white-fronted goose (*Anser albifrons*) and the willow grouse and ptarmigan (*Lagopus mutus, L. l. lagopus*). In some areas some species attain a high density, such as Middendorff's vole (*Microtus middendorffii*), the narrow-skulled vole (*M. gregalis*), the Norway lemming (*Lemmus lemmus*), the northern red-backed vole (*Clethrionomys rutilus*), the long-tailed arctic ground squirrel (*Citellus parryi*), the blue goose (*Chen caerulescens*), the brent goose (*Branta bernicla*), the red-breasted goose (*B. ruficollis*), the barnacle goose (*B. leucopsis*) and the whooper and Bewick's swans (*Cygnus cygnus, C. bewickii*), etc.

The musk ox (*Ovibos moschatus*), one of the most highly specialized species of arctic mammal, appeared there comparatively recently. It had actually been almost exterminated by the ancient people (on Taymyr it is possible to find a well protected wild herd). Now it is again being

Fig. 50. The bean or seed goose (*Anser fabalis*), a typically herbivorous tundra bird (photo.: A. V. Krechmar).

reintroduced from America and established in Taymyr and on Ostrov Vrangelya. But one of the mightiest of the herbivores of the tundra, the mammoth, which was alive up to less than 10000 years ago, has been completely eradicated.

At present there are within the tundra zone a few seed- and fruit-eating mammals such as the wood mouse (*Apodemus sylvaticus*), squirrels (*Sciurus vulgaris*) and the Eurasian chipmunk (*Tamias sibiricus*). Birds consuming seeds and fruits are comparatively well represented. True enough, some typical species of this kind inhabit in general only the southern areas of the zone. The most specialized seed-eaters, the redpolls (*Acanthis flammea*, *A. hornemannii*) are mass inhabitants of the forest tundra and the shrub-tundra. The races of the redpolls, the common (*A. flammea*) and the hoary (*A. hornemannii*), are difficult to distinguish and there seems to be little reason for the distinction. During the summer they also feed on insects, but for the major part of the year they consume seeds of various bushes and grasses. As nestlings they are already fed partly on seeds. Representatives of the southern Eurasian tundra such as the little bunting and Pallas's reed bunting (*Emberiza pusilla*, *E. pallasii*) are less specialized seed-eaters.

Small birds with a mixed food pattern are the most numerous. They consume both seeds and live food, that is insects. The following three species are very characteristic representatives of the arctic fauna, the Lapland bunting (*Calcarius lapponicus*), the shore lark (*Eremophila alpestris*) and the snow bunting (*Plectrophenax nivalis*). They belong to monotypic genera.

The Lapland bunting and the snow bunting do not occur outside the limits of the tundra and have no close relatives. This indicates that they have a long-term relationship with the tundra landscape and a high degree of adaptation to conditions there. During summer these species feed mainly on insects which are also fed to their nestlings. During autumn and often also in summer during cloudy weather they switch to feeding on the seeds of sedges (*Carex*), rushes (*Juncus*), saxifrages, etc.

Seeds, juicy berries and bulbils of viviparous species (*Bistorta vivipara*, some saxifrages, etc.) serve as an important supplementary food for snipe. The ruff (*Philomachus pugnax*) and plovers (the golden, lesser golden as well as the black-bellied or grey plover *Pluvialis apricaria*, *P. dominica*, *P. squatarola*) consume large quantities of the large seeds of Pallas' buttercup (*Ranunculus pallasii*), which grows in water. During autumn in the southern tundra, many birds feed on bog-whortleberries, cowberries (*Vaccinium uliginosum*, *V. vitis-idaea*) and cloudberries (*Rubus chamaemorus*), especially such granivorous gamebirds as willow grouse and ptarmigan (*Lagopus*) and also skuas (*Stercorarius*) and ducks (Anatinae). In the typical tundra the role of berries is negligible and in the arctic tundra there are none.

There is much that is peculiar and even paradoxical in the food relationship between vertebrates and tundra vegetation. All the typical herbivorous birds and mammals of the subarctic are in principle polytrophic. This means that they are able to consume not only plants but widely differing kinds of food. Thus, reindeer will eat mushrooms, lemmings, bird's eggs and so on, in addition to vegetable matter; lemmings will consume different grasses, twigs, leaves, roots, berries and flowers, ptarmigan will feed on leaves, berries, flowers, buds, etc. Not a single one of the indigenous inhabitants of the tundra have any morphological characteristics or physiological traits which could be said to be signs of any narrowly specialized feeding pattern (e.g., adaptations for the gathering of some special kind of food or for digesting it). Northern reindeer, lemmings and geese as well as ptarmigan are all herbivorous animals in the wide sense of that word. The snow hare (*Lepus timidus*) or typical seed-eaters such as the redpolls, (*Acanthis*) are more specialized herbivores, i.e., consumers of twiggy food. They are representatives of an essentially non-tundra fauna which inhabits mainly the southern parts of the subarctic.

However, in certain areas or at a certain time of the year there is, as a rule, a small number of foodstuffs which play a major role in the food pattern of each of the inhabitants of the subarctic. Thus, geese, both the bean goose and the white-fronted goose (*Anser fabalis*, *A. albifrons*), will consume very different kinds of vegetable matter, but mainly the green and juicy parts. One or two species of grasses, e.g., *Arctophila fulva*, and cotton-grasses (*Eriophorum angustifolium*, *E. vaginatum*), sedges (*Carex stans*), etc. are most often found in their stomachs. The favourite food of geese is the juicy portion of cotton-grasses and sedges just above the roots which they skillfully pull out of the moss cover. It is normally hard to spot where the geese have been feeding. However, in certain localities, where the cotton-grass and the sedges are especially vigorous and juicy, the vegetation may be badly torn up. After geese have fed, the surface is covered with remnants such as leaves which have been discarded after they have consumed the basal parts just above the root.

The willow grouse (*Lagopus l. lagopus*) in Taymyr in winter feeds almost exclusively on shoots and buds of willows but the ptarmigan (*L. mutus*) feeds on buds and catkins of alder (*Alnus*). During spring both species mainly consume leaves of *Dryas*.

Thus, the range of the basic food of typical herbivores is really quite limited. Many researchers, studying the food relationships of vertebrates in the subarctic, have come to the conclusion on the basis of these facts that vertebrates are more restricted to remarkably narrow food sources than the inhabitants of landscapes situated further south. For example, many authors consider the lemmings to be stenophagous, i.e., animals adapted to feeding very strictly on limited types of food. Such opinions are expressed in all the reviews of the ecology of tundra vertebrates (e.g., in papers by Dunayeva, 1948, Shvarts, 1963, Danilov, 1966 and Yudin, 1976, etc.).

However, this is not a question of food specialization nor is it stenophagy. When these terms are used, one usually has in mind an adaptation to feeding on a certain kind of food. Here we find essentially an evidence of polyphagy. This means nothing more than the ability to consume the kinds of food which at a given time are the most nourishing and most easily accessible. During winter, naturally, the food available for the ptarmigan consists of shoots of dwarf shrubs protruding above the snow. During spring, as the first to become free of snow, the young shoots of *Dryas* are consumed by these birds. A single goose, such as a young adult, may select patches in the tundra for feeding where there is an abundance of cotton-grass, but when sitting on eggs, geese consume only what is available close to the nest.

The reindeer switch from green summer feed to feeding on Reindeer moss (*Cladonia alpestris*) and twigs as food on their winter grounds. The Siberian lemming (*Lemmus sibiricus*) is able to eat the greatest range of plant types but the majority of this species live on nothing but the fleshy portions of sedges and cotton-grass (*Carex* and *Eriophorum*), the most numerous of all the higher plants in the moss cover, whereas the other species of herbs are hardly touched by these animals. In the polar desert, however, where there are no sedges or cotton-grass, the lemmings consume other grasses, saxifrages, chickweed (*Cerastium*), poppies (*Papaver*) and species of *Draba*.

The most abundant Passeriformes in the tundra, the Lapland bunting (*Calcarius lapponicus*), the snow bunting (*Plectrophenax nivalis*) and the shore lark (*Eremophila alpestris*), will in the course of a season consume mainly seeds and insects, sometimes one or the other, sometimes both. During spring they may also feed on rubbish at the settlements. It is characteristic of birds to switch to the kind of food which occurs in the greatest amounts. Thus, in the southern tundra at the time when blueberries and cowberries are ripe, these are consumed in great quantities by ducks (Anatinae), willow grouse and ptarmigan (*Lagopus*), skuas (*Stercorarius*) and geese, etc.

Certainly, forms of specialized herbivores are also well known. This is easily demonstrated when comparing related species. Thus, the Siberian lemming (*Lemmus sibiricus*) displays more constancy in its choice of food, depending heavily on that occurring abundantly in the tundra such as cotton-grasses and sedges. The food of the collared lemming (*Dicrostonyx torquatus*) is less variable. It often eats leaves of bushes and a variety of plants (*Salix, Betula, Vaccinium, Dryas, Rubus,* etc.). Besides, the collared lemming is known for its ability to store food for the winter in the same way as the pika (*Ochotona alpina*). This is particularly noticeable for animals living on Ostrov Vrangelya where in their 'pantries' over 50 different species of plants have been found (Yudin *et al.*, 1976).

Other examples of essentially different feeding habits are found in respect to the willow grouse and ptarmigan (*Lagopus l. lagopus, L. mutus*); see Table 9. Each season the nourishment of these species is basically different. During winter the willow grouse prefers twigs of the willow (*Salix*) while the ptarmigan takes the buds of alder (*Alnus*).

The monotonous type of food used by the herbivores on the tundra is largely determined by the poverty of the arctic flora. The herbivorous animals and birds consume everything that at a given moment is being produced most abundantly. This is neither stenophagy nor food specialization but one of the developmental processes leading toward potential polyphagy, i.e., the ability of utilizing a variation of food depending on season, weather and the stage in the life cycle of the basic food.

Phytophagous insects

Phytophagous insects occur in very different forms in the desert, the steppes and the forests. Many feed exclusively on particular parts of the plants: roots (rhizophages), wood (xylophages), leaves (phyllophages), flowers (anthophages) or on the nectar produced by them (nectariphages

Table 9. Comparison of food preferred by the willow grouse (*Lagopus l. lagopus*) and the ptarmigan (*Lagopus mutus*) in Taymyr (data according to Pavlov, 1974)

Food source	Willow grouse %	Ptarmigan %
Dryas leaves	12	10
Birch buds and leaves	13	7
Willow leaves, buds, twigs	44	2
Alder leaves, catkins	1	77
Seeds and berries	10	2

or anthophiles*), etc. There are no xylophages in the tundra zone, i.e., bark beetles, wood-borers or 'horntails' (Siricidae), in spite of the fact that some of them could actually live in the branches of the bushes. Consumers of herbaceous plants occur in great quantities at more southerly latitudes, for example the locusts, but reach only as far as the forest tundra. There are very few colony-forming social insects (aphids, scale mites) here.

Due to the severe living conditions in the soil, the number of rhizophagous insects is not very high. They are not as well known here as in other zonal associations. For example, larvae of the click beetles (fam. Elateridae) the so-called 'wire-worms', are found only as far as the southern border of the typical tundra subzone. The larvae of the pill beetle (*Byrrhus*), feeding only on roots, inhabit almost all the territories of the southern and the typical tundra but are nowhere very common. The most common rhizophages in the subarctic are the larvae of a weevil (*Lepyrus*) which feed on the roots of willows (*Salix*) and possibly also of other plants (see Fig. 51). On slopes covered by herbs and shrubs their numbers can amount to some dozens per square metre. Larvae of the pea weevil (*Sitona*) are often found on the roots of leguminous plants.

The most important group of subarctic rhizophages are the caterpillars

Fig. 51. Representatives of rhizophages within the tundra zone. 1, Larvae of the weevil genus *Lepyrus*; 2, caterpillar of the tortricid moth *Olethreutes*; 3, caterpillar of the tortricid moth *Epilemba*.

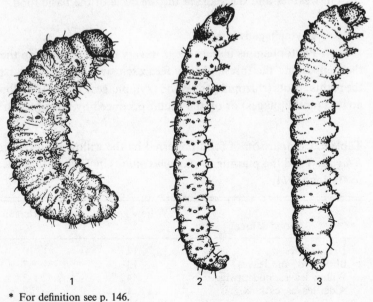

1   2   3

* For definition see p. 146.

of the tortricid moths. In the fleshy roots of species of louseworts (*Pedicularis oederi, P. dasyantha*) we frequently encounter caterpillars belonging to the genus *Olethreutes*. The development of these caterpillars is interrupted by a lengthy diapause. Therefore it is very difficult to hatch its imago stage and this makes it difficult to establish to which species they belong. Larvae of another genus of tortricid moths, *Epilemba*, are found in hollow roots. In Taymyr it is represented by *E. guentheri*.

The orange-coloured larvae of the fungus fly, which belong to the genus *Bradysia* of the family Sciaridae, are often met with on the roots of the lousewort (*Pedicularis*). To this genus also belong such well known agricultural pests as the cucumber fly, which, by the way, has been widely introduced into arctic villages in association with greenhouse cucumber crops.

On the watersheds in the moss tundra, especially where it is boggy, rhizophages are far less numerous. There we find the larvae of small Diptera – gnats and mosquitoes – and also of springtails. However, all these reside on dying or dead roots or rhizomes which, possibly, had previously been subjected to attack by nematodes and fungi.

It may seem paradoxical but the common rhizophages do enjoy some form of success in the arctic tundra subzone where it is colder and where there is no dense moss cover which makes the heating of the soil more difficult and so suppresses the growth of flowering plants. This kind of tundra is very beautiful; there are many herbs among which are species with various types of roots and rhizomes (saxifrages, *Draba, Pedicularis*, grasses, *Parrya, Lloydia*, etc.) There are, however, no specialized rhizophages in the soils there, such as the rhizophagous larvae of weevils or tortricid moths but a multitude of animals belonging to collective groups of saprophages using various remnants of plants as food, including both dead and living roots. Such are the larvae of, e.g., the craneflies (Tipulidae). Especially characteristic are *Tipula carinifrons, T. glaucocinerea* and *T. lionota*. Their numbers amount to hundreds per square metre.

The phyllophagous species among the phytophages are dominant in numbers in the tundra, just as in other zones. But here this group is poorly represented and in the arctic tundra their role as an important part of the biocenosis is in general almost none. The most characteristic trait of the tundra complex of phyllophages is the minimal role of sucking insects, feeding on plant juices with their biting probosces (aphids, scale mites, leaf hoppers, various bugs, etc.). Two to three species of bugs, leaf hoppers, aphids and cicadas inhabit the typical tundra subzone. They are absent in the arctic tundra subzone. Most of these insects occur in the southern tundra but even there they do not play any major role compared with that in forests, steppes and deserts.

They are able to reproduce in great quantities only on special plant species. Thus, on sagebrush (*Artemisia*) we can find quantities of mirid bugs belonging to the genus *Orthotylus*. On shrub species of willows, especially on the catkins of *Salix reptans* and *S. lanata*, there is an abundance of leaf hoppers belonging to the genus *Psylla*, especially *P. palmeni* and *P. zaicevi*. Locally in grassy habitats there are many small 'cicadas'. In the Taymyr they are represented by *Streptanus arctous*, *Hardya youngi* and *H. taimyrica*. On the basal parts of plants we may find colonies of aphids just above the roots, and in thick moss sward large discoid flatworms, *Arctorthezia cataphracta*, which are widely distributed in the Arctic.

The depauperate number of sucking phytophagous insects here is not conditioned by their characteristic relationship to the food source but rather the opposite. Their mode of feeding and their life style (low mobility, colony formation and the ability of many species to reproduce parthenogenetically) are advantageous under the severe arctic conditions. Here, it is the matter of their development. These insects develop without complete metamorphosis, i.e., they have no stages of true larvae and pupae like the beetles, the butterflies and the flies. Small insects similar to the adults and usually consuming the same food as the parents emerge directly from the eggs. Thus, while still in the eggs the basic organs of the adult organism are perfected. This distinguishes them from insects with a complete metamorphosis, the larvae of which are entirely different from the adults. These have to grow and moult over a long period of time until they acquire the traits of the imago.

The eggs of insects with incomplete metamorphosis are supplied with a large amount of nourishment, as a result of which the embryo can pass through all the necessary stages. Much heat is necessary for the development of such eggs but there is not always enough of that in the tundra. Therefore the majority of the representatives of orders of insects with incomplete metamorphosis (cockroaches and other orthopteroids, Dermaptera, Psocoptera, Homoptera and Hemiptera) do not find conditions suiting them here. The eggs of insects with complete metamorphosis contain sparse nourishment. They need less heat for developing. The later development of the larvae hatching from the eggs and leading principally to another life form than the imago may happen quickly or slowly depending on weather, type of biotope, abundance of food, etc. Complete metamorphosis is more advanced and therefore better suited for the severe conditions of the arctic environment. The major part of the entomofauna within the tundra zones consists of insects developing with metamorphic stages, i.e., representatives of the orders of Diptera, Hymenoptera, Lepidoptera, Trichoptera and Coleoptera.

The tundra fauna of chrysomelid beetles is very rich. The majority belongs to the genus *Chrysolina* (recently in part renamed *Chrysomela*). Species of this genus reach far northwards, all the way into the arctic tundra. The most abundant chrysomelid in the Eurasian tundra is *Ch. septentrionalis*. The beetles and the larvae of this species feed on the leaves of various species of plants, e.g., buttercups (*Ranunculus*), larkspur (*Delphinium*), forget-me-nots (*Myosotis*) and louseworts (*Pedicularis*) and many others. They also devour the buds. The larvae grow very slowly, requiring 2–4 years for maturation while wintering over at different ages. In Eurasia we also encounter some arctic species belonging to this genus, *Ch. subsulcata*, *Ch. tolli* and *Ch. bungei*. These chrysomelids inhabit local biotopes in the southern tundra, mainly herb-covered slopes, but in the arctic tundra they come out on to the watersheds and colonize rather zonal associations which are affiliated with the abundance and variety of the vegetation of flowering herbs in that type of landscape.

Some chrysomelid species enter the tundra from the south, feeding on the leaves of shrubby willows. Most common among them are *Phratora polaris* and *Phytodecta pallida*. Both these species occur abundantly in the forest tundra and the southern tundra.

In wooded areas big and brightly coloured chrysomelids are well known which were recently still referred to the genus *Melasoma* but now have been transferred to *Chrysomela* (alder, aspen and poplar chrysomelids, etc.). For a long time species of this genus were not known from the tundra and only in the most recent time have some species been described from Taymyr, north-eastern USSR and Alaska. In Taymyr *Ch. taimyrensis* inhabits the typical tundra belt. Larvae of this species develop mainly on arctic willows (*Salix arctica*) and have a striped coloration. When raised under laboratory conditions, the larvae are also able to feed on other local willows: *S. reptans*, *S. pulchra*, *S. lanata* and *S. polaris*. Fully viable beetles of these species develop on all these willows but the larvae do perish on one species with small and coarse leaves (*S. nummularia*). Thus, these larvae are facultative monophages, preferring one kind of food but able to survive on several. Locally their numbers are high; the larvae are able to devour up to 50% of the leaf mass. When the larvae are agitated they secrete droplets of a whitish fluid with a sharp smell of camphor from the sides of the body. In calm weather it can be detected at a considerable distance, and thus it is possible to locate the presence of the larvae.

In contrast to the typical tundra chrysomelids belonging to the genus *Chrysolina*, species of *Chrysomela* develop very fast, usually in the course of a single year. The larvae feed intensively and complete their growth by the middle of August. Only adults of this genus winter over. The life cycle of this genus must definitely be completed within a single summer. Thus, the type of development reflects a southern provenance of this genus. Under the severe conditions of arctic areas this is a disadvantage. Indeed, because of this, chrysomelids do not extend as far into the tundra zone as species of the genus *Chrysolina*.

The most important phyllophages are the larvae of sawflies. In the moss tundra they constitute almost half of all the inhabitants on the herbage, while in herbaceous meadow-like sites almost one quarter. The majority belong to the subfamily Nematinae and have an affinity for willows. Almost all the willow species on the tundra are affected by galls, produced by sawflies (see Fig. 52, 1–3). The galls (a thick, fleshy growth on the leaves in which the larvae live) are especially common on the shrubby willows (*Salix reptans, S. lanata*), less so on the creeping *S. arctica, S. pulchra* and *S. polaris*. In Taymyr the galls on all the willows are produced by a single species of sawfly, *Pontania crassipes*. It is widely distributed in the territories of northern Europe and Siberia. In the tundra it develops over a single season. It winters over as a pre-pupal stage inside the gall which together with the leaf falls to the surface of the ground. Many other species of tundra sawflies live on leaves without forming galls.

The larvae of the sawflies are often taken for true caterpillars. They are definitely similar to butterfly or moth caterpillars and lead a similar life devouring green parts of plants. Among ecologically similar groups of organisms a reverse correlation is often observed: where there are many species belonging to one group there are few of another. In the steppes, caterpillars predominate, in the broad-leaved forests they are plentiful, but there are also many sawflies and in the tundra the latter predominate. They are especially notable on the typical tundra, where the number and specific variety of sawflies is especially high; there are few species of the butterfly family.

Information on the food relationships of butterfly caterpillars in the tundra zone is very scanty. For example, almost nothing is known regarding the food of the most abundant of the arctic species belonging to the diurnal Lepidoptera, the 'mother-of-pearl' genus *Brenthis*. A species of banded alpine butterfly, *Erebia fasciata*, actually associated with grasses is indigenous in the northern half of the subarctic. Also supposed to develop on grasses is the grayling (*Oeneis ammon*), rare in the Siberian and European tundras. On various leguminous species we can find caterpillars of the yellow butterflies *Colias nastes* and *C. hecla*.

There are many so-called 'nocturnal' Lepidoptera or moths in the

Fig. 52. Typical insect pests on plants in Taymyr. 1–3, Galls on willow
leaves made by the sawfly *Pontania crassipes*; 4–5, catkins of
arctic and creeping willows (*Salix arctica, S. reptans*) damaged by
the larvae of the sawfly *Pontopristia*; 6, leaves of arctic willow (*Salix
arctica*) damaged by larvae of the leaf cutter *Chrysomela taimyrensis*;
7, roots of *Pedicularis oederi* (Oeder's lousewort) damaged by larvae
belonging to the fly family Muscidae; 8, roots of another lousewort,
(*P. dasyantha*) damaged by caterpillars of the tortricid moth
Olethreutes; 9, grass spike of arctic Poa (*Poa arctica*) damaged by the
carabid beetle *Amara alpina*.

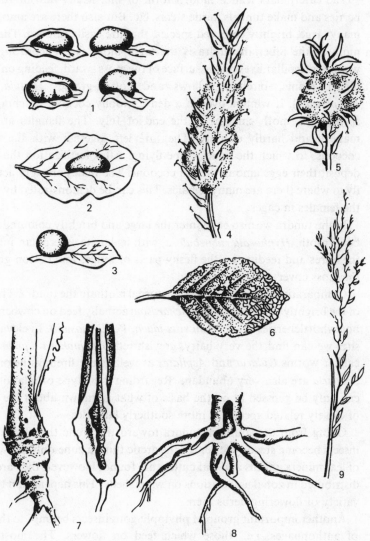

tundra: the night moth *Polia richardsoni*, the noctuid moth *Graphiphora liquidaria*, the moths of the Pyralidae, the 'woolly bear moth' (*Hyphoraia subnebulosa*), the 'inchworms' (fam. Lymantridae), etc. On the whole, the specific composition of the population is rather well known, but their ecology is almost unknown. There is almost no information on what plants the caterpillars consume, what kind of damage they do or what distribution they have. The overwhelming majority of the moths are small and obscure forms, flying badly and only for short periods of time.

The caterpillars lead a hidden form of life. Many devour roots and berries and make tunnels inside stems, etc. But also there are among them many large, brightly coloured species, that are easily noticed. The largest moth in the Siberian tundra is the night moth *Gynaephora lugens*. Its woolly caterpillar lives on the surface of the moss sward, feeding on various higher plants, birches, willows and cotton-grasses (*Betula*, *Salix*, *Eriophorum*). It winters over in a dense, woolly cocoon, hatching in the spring. The moth emerges at the end of July. The females are rather massive and hardly ever fly. They are left together with the rows of cocoons, to which the males come flying. After fertilization the females deposit their eggs among the old cocoons. The males are very active and fly to where there are many females. This can be demonstrated by keeping the females in cages.

In the tundra we also encounter the large and brightly coloured 'woolly bear' moth, *Hyphoraia subnebulosa*, with its lanate caterpillar, inhabiting the mires and feeding on the fleshy parts of sedges and cotton-grasses in the moss cover.

Comparatively there are many nocturnal moths in the tundra. The larvae of the brightly variegated *Polia richardsoni* actually feed on cowberries and bog whortleberries (*Vaccinium vitis-idaea*, *V. uliginosum*). In elevated, dry sites we can find the very hairy, greyish-red *Graphiphora liquidaria*. The canker worms *Cidaria* and *Aspilates* as well as the firebugs belonging to *Scoparia* are also very abundant. Regarding their type of relationship, it can only be guessed at on the basis of what is known about the biology of closely related species at more southerly latitudes.

Going from the southern tundra toward the arctic the phytophagous insects become steadily fewer. In the arctic tundra zone only a few species of leaf miners, moths and leaf cutters are found. However, they are widely distributed in zonal associations on watersheds. This depends on the great variety of flowering herbs there.

Another important group of phytophagous insects belongs to the group of anthophages, i.e., those which feed on flowers. The most typical representatives of this group are the leaf cutter genus *Hydrothassa*, of

Fig. 53. Representatives of anthophilic and anthophagous insects in the typical tundra subzone. 1, Fly belonging to the family Muscidae on the flowers of *Dryas*; 2, fly of the family Empididae on the flower of *Dryas*; 3, male *Aedes* mosquito drinking nectar from the flower of a buttercup (*Ranunculus*); 4, the chrysomelid beetle (*Hydrothassa hannoverana*, devouring the corolla of a buttercup (*Ranunculus*); 5–6, the carabid beetle *Amara alpina* on the flowers of a forget-me-not (*Myosotis*) and a cinquefoil (*Potentilla*).

which *H. hannoverana* reaches the farthest north of all. This species lives at middle latitudes on the marsh marigolds (*Caltha*), but in the tundra also on buttercups (*Ranunculus*) e.g., *R. borealis* and *R. sulphureus*. Its larvae feed on petals, anthers and ovaries of the flowers (Figs. 53, 4).

The weevil *Phytonomus ornatus*, another anthophage, belongs here, too. It is distributed from Mongolia into the Siberian tundra. Its peculiar larvae, more similar to butterfly caterpillars than to larvae of other weevils, develop on milk vetch and oxytrope (*Astragalus, Oxytropis*), feeding on the flowers while still in bud, and in the course of anthesis gradually changing over to feeding on the fruits. Locally they damage as much as 20% of the milk vetch and oxytrope flowers. The larvae pupate in strong and semi-transparent, woven cocoons at the end of the summer at the time when the fruits are maturing. Small orange midge larvae are also associated with leguminous flowers. They inhabit corollas of the arctic milk-vetch (*Astragalus umbellatus*) in especially large numbers. In some years all the flowers of these plants are affected.

The little beetle *Amara alpina* occurs in great numbers in many areas. During the hours of night when light intensity is low but the temperature still quite high (not below 8 °C), the adults climb up on different plants and devour the tops of the flowering shoots, buds, flowers and young fruits (see Fig. 53). They feed especially intensively on the panicles of the grass *Poa arctica*. It is often possible to observe several specimens at once on the same panicle. They also consume the anthers of other common grasses such as *Alopecurus alpinus*, and can be found among the inflorescences of the only umbellifer occurring at high latitudes, i.e., *Pachypleurum alpinum*. In Taymyr it is easy to find more than 20 species of plants, the inflorescences of which have been eaten by *Amara alpina*.

The carpophages, i.e., those devouring fruits and seeds, may be included among the anthophages as well. Previously representatives of anthophages, e.g., the weevil *Phytonomus ornatus* and the little beetle *Amara alpina*, have been mentioned as also feeding on ovaries and mature seeds. They can to some extent be referred to the group of carpophages as well. There are also many carpophages among the small gnats. Their larvae develop in the fruits of a few dozen species of tundra plants. Fruits of the lousewort (*Pedicularis*) are very strongly affected, especially those of the two species growing in mires in the typical tundra subzone, *P. sudetica* and *P. hirsuta*. The fruits of one of the typical arctic leguminous species, the blackish oxytrope *Oxytropis nigrescens*, which grows in long rows over gravelly sites, are eaten by larvae of the small nocturnal Hymenoptera, close relative of the moths (the description of this group needs specialized research). The fruits of different willows, legumes, louseworts and composites are also consumed by the larvae of true flies. The sawwort, *Saussurea tilesii*, is especially badly affected. Unfortunately and in spite of repeated attempts, I have not succeeded in raising the imago of their larvae.

Fruits of tundra willows are badly damaged by leaf miners belonging to the genus *Pontopristia*. The affected catkins are easy to recognize because of the dense 'pillows of fluff' formed where the larvae are present (see Fig. 52, 4, 5). In Taymyr it is possible to raise the imago from larvae on the catkins of four species of willows. On each of them a particular species of leaf miner can be identified. This indicates great specialization among the carpophages in comparison with the phyllophages. However, the leaf miners mentioned above were able to develop also on many species of willows.

There are very many carpophages among weevils within the forest belt. Only a few species are met with in the tundra zone. Besides *Phytonomus ornatus* associated with seeds of legumes, some other carpophagous species have recently been described: *Apion tschernovi* and *A. wrangelianum*. In the southern parts of the tundra zone catkin weevils belonging to the genus *Dorytomus*, e.g., *D. flavipes*, *D. nordenskioldi* and *D. salicinus* are found on willows in the European tundra.

All the trophic groups of phytophagous insects are subjected to progressive impoverishment proceeding northwards from the southern limits of the subarctic. In some cases the number of species groups is sharply lowered independently of the presence of food plants. As an example may be mentioned sucking insects, locusts and many Lepidoptera. The reasons for the impoverishment of these groups may be insufficient heat, the brevity of the summer season, the lack of necessary photoperiodic alternation between night and day, etc. The distribution of phytophages is also limited by access to food plants (with respect to xylophilous insects, or phyllophagous species associated with shrubby willows, etc.).

At the same time a sharp difference can be observed in the tundra zone between the phytophages and their food plants. Often a conflict arises there in their distribution which does not always depend on zonal borders. This is especially true in the case where plants and the insects associated with them inhabit intrazonal biotopes, e.g., mires or river banks. A very clear case is represented by the leaf cutter *Hydrothassa hannoverana*. It is distributed over an enormous area from the forest tundra into the arctic tundra and exists on the marsh marigold (*Caltha*). Botanists distinguish two species of marsh marigold within the tundra zone: *C. palustris* in the southern areas and *C. arctica* reaching farther north. The leaf cutter does not stay only on these two species on the tundra occurring there also in greater numbers than in the forest belt, but in addition, it is able to switch over to feeding also on various species of *Ranunculus*. This example belongs to the case where a species at the limit of its distribution range makes use of all its ecological opportunities.

Some other cases are represented by the distribution and food association of leaf cutters such as *Chrysolina marginata*. This species inhabits an area extending from the tundra to the steppes and semi-deserts and it is associated with different species of composites, e.g., *Anthemis*, *Alchemilla*, etc. In the tundra zone their food association is narrower. For instance, in Taymyr they are met with mainly on sagebrush species (*Artemisia*). In the arctic tundra subzone where these plants are less numerous, this leaf cutter is not found.

Chrysolina marginata is an example of an oligophage, i.e., a species associated with plants belonging to a single group (in this case the Compositae). Other typical oligophages are the weevils distributed on legumes, the leaf miners which attack willows in Taymyr, some leaf cutters, etc. All these are extra-arctic in origin. Only oligophages have spread more intensely into the tundra zone from the south but this is, of course, possible only where they can find plants to fit them.

Narrowly specialized monophages, associated with a single species of food plant together with the polyphages which feed on different species, penetrate much less vigorously into the tundra. The most typical polyphages widely distributed within the forest belt (e.g., the locusts, different bugs, Lepidoptera and leaf cutters) are limited in their range to the northern forest tundra or the northern taiga.

There are also polyphages among the insects in the typical tundra subzone, e.g., leaf beetles belonging to the genus *Chrysolina*, those described above belonging to the family Lymantridae (*Gynaephora*), and some Lepidoptera. Among the monophages there we find probably all the species of the willow catkin sawfly (*Pontopristia*). In addition there are also facultative monophages which are in principle able to feed on many plants, but exist mainly on a single species. As an example may be mentioned the leaf cutter *Chrysomela taimyrensis*. The problem concerning gradually specializing phytophagous insects on the tundra is very interesting and important, not least in connection with the possibility of them becoming pests affecting the development of agriculture.

The majority of the oligophages are associated with the more important groups of tundra plants such as willows, legumes, scrophularias, composites and ranunculaceous species. The relative number of insect species feeding on these plants is not smaller but in a number of cases actually higher than at more southerly latitudes. A very rich complex lives, for example, on the willows. This comprises sawflies, catkin and root-eating weevils, leaf cutters, moth caterpillars, Diptera and pollinating bumble-bees feeding on pollen and nectar, etc. Many insects are able to exist at the expense of the legumes, e.g., carpophagous and rhizophagous pea weevils, moth and butterfly caterpillars, and larvae of flies, developing inside the fruits, sawbugs, bumble-bees, etc.

At the same time the entomofauna of such important groups of plants as crucifers and grasses is very poor. At more southerly latitudes crucifers are associated with an enormous number of insects, cabbage butterflies, firebugs, noctuid moths, different sawflies and leaf cutters and especially with the variable soil fleas (*Haltica*). In the tundra, species of these plants are abundant and highly variable (many genera such as *Draba, Parrya, Cardamine, Braya, Cochlearia, Thlaspi* belong there) but they house few insects. Their flowers are not regularly visited by anthophages. In the southern tundra subzone and in part in the typical tundra subzone some examples of white moths are met which develop on crucifers, especially the cabbage moth (*Pieris napi*) and the white alpine moth (*Synchloe callidice*). They are, however, few in number and not typical of the tundra. The lack of an entomofauna specialized to feed on crucifers is especially noticeable in the arctic subzone where these plants are extremely abundant.

The richest fauna attacking grass species in the northern belt of the subarctic is associated with only a few species growing mainly in its southern areas. These are grass borers, the larvae of which live inside the stems and also some species of caterpillars devouring the leaves. There is almost no information on the insect fauna associated with the other major group of monocotyledonous plants of the tundra, the sedges. Much indicates that it is extremely poor.

Problems regarding the amount of activity of phytophagous insects under arctic conditions and the degree of damage done to the plants by them are interesting. Information thereon is, however, limited. It should be noted that the damage produced by phytophagous insects to tundra plant associations is exceeded many times over by that caused by lemmings, which appear to be the main consumers of the vegetation. They have such an extensive effect on it that all other damage appears unimportant. The insects are able to rival the lemmings in respect to the amount of plant material consumed only within special biotopes. Thus, in southern parts of the tundra zone the sawflies and leaf beetles and their larvae consume about 12% of the leaf material in willow brush. They affect legumes very badly, especially the flowers and fruits. Larvae of weevils often consume up to 70% of the fruits of *Oxytropis*.

These problems need to be investigated much more thoroughly. Cultivation is spreading into the Arctic; agriculture has long since been introduced to the northlands. In the near future there will be problems regarding research and practical work on regulating the number of phytophagous insects. It is therefore especially important to acquire information on the behaviour of insects and what their relationships are with the plants under natural conditions in subarctic environment as yet untouched by man.

Insect pollinators and flowers

The anthophilous insects, that is those which visit flowers for nectar and take part in cross pollination, must be distinguished from the anthophages described above. Plants pollinated by insects are called entomophilous. (The words 'anthophile' and 'entomophile' are derived from the Greek words for flower, *anthos*; insect, *entomon* and love, *philos*.) The interdependence of anthophilous insects and entomophilous plants is a classic example of symbiosis, which has an enormous impact on the evolution of both groups of organisms.

Long ago the pollination of plants under arctic conditions started to attract the interest of scientists. Already by the end of the nineteenth century the great botanists of that period, E. Warming, O. Ekstam and F. Kjellman, published results of investigations on pollination of tundra plants (a review of such works is given in papers by Shamurin, 1966 and by Chernov, 1966). Also the unique specialities of entomophilous plants in the Arctic were exposed. It was noted that many, in spite of having big and brightly coloured flowers, were able to develop seeds without cross-pollination or that they had become adapted to autogamy (i.e., auto- or self-pollination). The switch to self-pollination is usually associated with a small supply of pollinators or with unfavourable conditions when these are active. Autogamy is reflected in the structure of the flowers which are simplified and have lost the characteristics of entomophily, that is, the size of their petals is reduced, the brightness of colour and the amount of nectar are also less, and so on. However, many plants which are found to be, or are believed to be, autogamous also have flowers with a typically entomophilous appearance. This is especially evident in the case of the louseworts (*Pedicularis*), one of the many important plant genera in the arctic flora. Their large, brightly coloured flowers are assembled into dense inflorescences, looking even more attractive to anthophiles. In spite of that the flowers have been found to be obligatory self-pollinators. It can be stated that this may be a result of the youthfulness of the arctic flora: while becoming autogamous the plants have not yet had time to lose their more advanced characteristics of entomophily.

Among the entomophiles, the most interesting are the plants with flowers structured so that they are able to be visited and pollinated only by particular insects which have become adapted for this purpose. As typical examples of such plants in the flora of the temperate belts, species like the sage (*Salvia*), the toadflax (*Linaria*) and the pea (*Lathyrus*) can be mentioned. Their nectar is located at the base of a cavity, in pockets or in a spur, and the positions of the anthers are adjusted according to the size of the pollinating insects. Usually these are bumble-bees, bees, large flies or butterflies with long probosces.

Such entomophily is represented in the arctic flora mainly in species belonging to three genera of the family Leguminosae (i.e., *Astragalus*, *Oxytropis* and *Hedysarum*) and the genus *Pedicularis* of the Scrophular-iaceae. Among all the examples known in the tundra zone, 15 belong to the leguminous species and 11 to the louseworts. A representative of the family Ranunculaceae, the larkspur of the genus *Delphinium*, is represented in the tundra flora by some species and forms among which *D. middendorffii* is the most common. Species of gentians (*Gentiana*) may also be included in this group. They have long trumpet-shaped flowers and are commonest in the eastern part of the Eurasian tundra.

The majority of entomophilous plants have flowers which can be pollinated by different insects. Such flowers are usually not very deep, have free-standing petals and easily accessible stamens, stigma and nectary (the buttercup, *Ranunculus*, is a typical example of such a flower). Nectar and pollen may be consumed totally by various insects, both by typical anthophages such as bumble-bees, or by different beetles, mosquitos, flies or deer flies. Such plants could be called non-specialized entomophiles. In the tundra rosaceous, ranunculaceous, composite, caryophyllaceous, saxifrageous and cruciferous species predominate among them. Such families as Papaveraceae, Valerianaceae, Crassulaceae, Boraginaceae, Polemoniaceae and Liliaceae usually have numerous single flowers.

In order to understand the peculiar interrelationship between insects and flowers, it is important to know the number of entomophiles within a flora and the number of species which need to be pollinated. Below we will consider the composition of the floras within some of the tundra zones from this point of view.

In the typical tundra zone on Taymyr in the neighbourhood of the research station Tareya, 236 species of flowering plants have been listed (Polozova and Tikhomirov, 1971). Of these, 60 species are grasses, sedges (*Carex*) and rushes (*Juncus*) in which insect pollination plays no role at all. Among the remaining 176 species, 19 may be considered as rather sparse in the flora, 11 are very rare and nine are representatives of a group for which there is no reason to suggest entomophily (e.g., birch, *Betula*; sagebrush, *Artemisia*; and species of Polygonaceae). After deducting these species, 137 remain within the area which potentially belong to an entomophilous flora. Of these, 86 species are relatively frequent and play a major role in the vegetation cover. However, we find less than 20 specialized entomophilous species in this area: eight belong to the Leguminosae, eight to the genus *Pedicularis*, one is a *Delphinium* and one a *Gentiana* (= *Comastoma*). All the rest are representatives of entomophilous herbs with non-specialized flowers.

Within this same area the associations of pollinators have also been studied (Chernov, 1978). In the neighbourhood of the Tareya research station we have found 140–150 species of insects visiting flowering plants. Of these about 90–100 species show the typical characteristics of potential pollinators (they consume nectar or pollen, are relatively active fliers, have relatively large and furry bodies). The really important ones with respect to pollination of the most typical entomophiles in the local flora belong to no more than two dozen species: three species of bumble-bees (*Bombus hyperboreus*, *B. pyrrhopygus* and *B. pleuralis*), 5–6 species of hover-flies (*Conosyrphus tolli*, *Heliophilus groenlandicus*, *Syrphus tarsatus*, *S. dryadis* and *Platychirus hirtipes*), one species of bluebottle or blue carrion fly (*Boreellus atriceps*), five or six species of the families of true flies, flower weevils and damsel flies, two to three species of moths and three to four species of sawflies (among which is, first and foremost, *Tenthredo*).

Evidently the general number of entomophilous plants in the typical tundra is approximately equal to the number of anthophilous insects. In the arctic tundra this correlation is slanted in favour of the entomophilous plants. Thus, in the environment around Dikson (on the eastern bank of the mouth of the Yenisey) we find only one species of bumble-bee and two species of large hover-flies but about ten species of plants which can be pollinated only by these insects. In more southerly zones the anthophilous fauna is definitely richer than the entomophilous flora. In coniferous forests it is possible to meet with some dozens of hoverfly species on a single specimen of sorrel (*Oxalis acetosella*). In the steppe meadows the species of bees and bumble-bees are considerably more numerous than the number of entomophilous plants.

Many of the peculiarities with respect to entomophily in the tundra zone are determined by such correlations, in particular by the development of competition among the plants for the insects available. As a result of such competition we can mention some peculiarities in the biology of many subarctic plant species, such as the capacity for growing in long rows or in the form of cushions or for forming swards with a concentration of great masses of flowers. No less important are flowering times, some detail of floral morphology to attract insects, scent, etc. All these problems need more research.

The manner in which entomophilous plants attract insects is very variable. This can easily be observed in the presence of bumble-bees on different plant species in a single locality (see Table 10). Research has revealed that the most attractive flowers are those of Middendorff's and of Adam's oxytropes (*Oxytropis middendorffii, O. adamsii*). These plants carry large and brightly coloured flowers assembled into head-like inflorecences. They exude much nectar. During a couple of days of good weather bumble-bees visited the majority of the flowers of these species. This assured, of course, successful pollination. At the same time it was evident that while these species attracted the majority of the bumble-bees, this reduced the opportunities for other plants to be pollinated.

Compact growth and concentration of flowers are other forms of adaptation directed toward attracting pollinators. However, a very large number of plants of the same species covering a wide area may also be of importance as an 'exhausting factor', provoking the bees to switch over to other species of plants. Thus, a small investigation revealed that a milk-vetch (*Astragalus subpolaris*) was the most abundant legumin-ous plant and a lousewort (*Pedicularis verticillata*) the most abundant scrophulariaceous one. Both species are typically entomophilous, but it

Table 10. Frequency of bees visiting flowers in localities with herbaceous meadow-like vegetation, Taymyr, 27 July 1969

Plant species	No. of flowers under observation	Visited by bees, %	Minimum no. of days impossible for bees to visit the flowers
Arctic milk-vetch (*Astragalus subpolaris*)	35475	10	10
Adam's oxytrope (*Oxytropis adamsii*)	1593	24	4
Middendorff's oxytrope (*O. middendorffii*)	770	96	2
Verticillate lousewort (*Pedicularis verticillata*)	6601	6	17
Oeder's lousewort (*P. oederi*)	640	14	7
Middendorff's larkspur (*Delphinium middendorffii*)	35	34	3

was just these species which were less visited by the bumble-bees than all the other species belonging to these families within this area.

The biotopic distribution plays a major role in the interrelationship between anthophilous insects and entomophilous plants. The bulk of anthophiles and entomophiles belong to local biotopes occupying small areas such as slopes beside a river, lake banks, gullies etc. This contributes to an increase in the efficiency of plant pollination there. In some areas with watershed tundra there are few pollinators. Mainly fairly small flies live there and often no bumble-bees at all. The lack of insects is one of the obstacles to full productivity of entomophilous herbaceous associations in moss–tundra plakors. We were able to convince ourselves of this by means of some experiments. Typical entomophiles, such as those belonging to the Leguminosae which are pollinated by bumble-bees and large hover-flies, were dug from the soil and transplanted while still in bud from river banks or slopes along gullies into watershed moss tundra a distance of 2–3 km away. Counting the number of seeds formed after anthesis showed that many flowers had not been pollinated. In any case, considerably fewer fruits were formed than in the biotopes from where these plants had been brought.

The propagation of leguminous species on the tundra would be impossible without the participation of insects. The main pollinators are bumble-bees and large hover-flies. In the typical tundra there are three species of hover-fly which are able to feed on pollen and nectar of the leguminous flowers. These are the large and brightly coloured *Conosyrphus tolli*, *Heliophilus borealis* and *H. groenlandicus*.

Concerning information on insects with respect to the pollination of leguminous plants we can also refer to the results of some experiments made when isolating their flowers from the anthophiles (see Table 11). The leguminous species are typically entomophilous. When isolated from insects they produce hardly any seeds. Seeds are, however, occasionally found on isolated oxytropes (*Oxytropis middendorffii*). It must be assumed that this species, being the most specialized and the one best adapted to pollination, almost exclusively, by large bumble-bees, has preserved – or developed – an ability for self-pollination which can take over when entomophily fails.

As already mentioned the arctic species of lousewort (*Pedicularis*) are often characterized by autogamy. The outward appearance of some of these species may already have convinced us that they belong to the entomophiles. Thus, *P. hirsuta* has small flowers with wide-open corollas which are covered by hairs and adpressed to the stem. When the flowers are isolated from insects, seeds are formed in the same number as when not isolated. Therefore it seems evident that this is a typically autogamous species.

The pollination process is fairly similar for the majority of the species of *Pedicularis*. When isolated a large number of fruits will set but perhaps in somewhat lesser quantity than for the non-isolated specimens. Furthermore, the fruits formed after isolation, that is, supposedly developed as a result of autogamy, reached a smaller size and were of definitely lower quality. Thus, *P. sudetica* produced on average 29 fruits when isolated, while the controls carried 62. This species grows in polygonal mires far from the herbaceous slopes with their concentration of pollinators. Its autogamy may be very significant for the preservation of seed production during years unfavourable for insect pollination. The hover-fly, *Conosyrphus tolli*, the larvae of which develop in the gullies of mires, is the most important pollinator.

The pollination of *P. capitata* has acquired a peculiar form. This species grows in small groups, certainly associated with the type of parasitism typical of the family Scrophulariaceae. Its flowers look distinctly entomophilous and are of large size but insects visit them comparatively rarely. However, isolation of the flowers has demonstrated that cross pollination

Table 11. Results of isolating flowers, western Taymyr (1969)

Plant species	Number of flowers isolated	Number of seeds set after isolation	Seed set on controls, %
Blackish oxytrope (*Oxytropis nigrescens*)	219	0	93
Adam's oxytrope (*O. adamsii*)	119	0	—
Middendorff's oxytrope (*O. middendorffii*)	18	1	71
Arctic milk-vetch (*Astragalus subpolaris*)	220	0	68
Alpine hedysarum (*Hedysarum alpinum*)	112	2	67
Oeder's lousewort (*Pedicularis oederi*)	120	2	64
Sudetian lousewort (*P. sudetica*)	127	15	83
Capitate lousewort (*P. capitata*)	83	12	—
Verticillate lousewort (*P. verticillata*)	73	4	79
Middendorff's larkspur (*Delphinium middendorffii*)	29	0	92

is of great importance to them. Thus, when isolated only 14% of the fruits will set. During anthesis the majority of the flowers have their stigmas and anthers arranged in full agreement with entomophily. However, during anthesis some flowers of this lousewort also thrust forward long stamens and styles from the corolla (see Fig. 54, parts 2 and 3) and these flowers display the typical traits of anemophily (i.e., wind pollination). There are some specimens where, after a period of flowering, all four stamens begin to protrude on long filaments and shed their pollen. In such cases, the stigma of the same flower will already have dried up which may indicate an adaptation for the prevention of self-pollination. All this points to the fact that for the capitate forms of the lousewort, wind dispersal has an essential role as an alternative method for the distribution of its pollen.

A similar capacity (i.e. exertion of stigmas and stamens and shedding the pollen for wind dispersal) is found also among the florets of *Pedicularis oederi*. The characteristics of wind pollination are usually not displayed immediately after the flowers open but some time after when no pollen has been deposited by insects. Anemophily is actually an alternative method, together with the basic entomophily, for assuring pollination. However, there are many indications that these species have also become adapted to some degree of self-pollination.

Thus, in the case of *Pedicularis* the basic peculiarities of the floral

Fig. 54. Flowers of the capitate lousewort (*Pedicularis capitata*). 1, Section through flower with a typical, entomophilous appearance; 2, flower with strongly elongated and protruding pistil; 3, flower with twisting stigma, protruding stamens and pollen-bearing anthers.

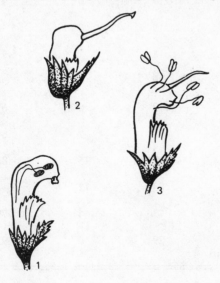

biology can be said to indicate that there, combinations of various types of pollination exist which have indeed prepared the flowers of this group of plants for the severe conditions within the tundra zone. In general the combination of different mutually supportive systems of adaptation for pollination is characteristic of many plants within the different zones. Kerner stated as early as 1903: 'there is already another function in store in case damage is done to the principal one'. Insect pollination of tundra species belonging to this genus is undoubtedly of great importance. But in Taymyr out of the six most common species of *Pedicularis* within the typical tundra subzone, four are predominantly entomophilous. Autogamous pollination produces only a small quantity of seeds.

All this forces us to admit that the tendency of the arctic *Pedicularis* towards autogamy has evidently been exaggerated in the literature. Most definitely a series of data bear witness to the opposite and to the fact that only in the subarctic entomophily has developed more strongly by comparison with latitudes farther south. Together with the autogamous species, entomophilous ones such as *P. oederi, P. adamsii, P. sudetica* and *P. verticillata* also occur. The brightly coloured and large flowers always attract insects among which are also some bumble-bees. Another hypothesis indicates a close association with *Conosyrphus tolli*, one of the most typical endemics within the tundra zone. It may be suggested that there has been a parallel evolution between this hover-fly and the *Pedicularis* species of the mires.

Autogamy has also been demonstrated for many species of supposedly entomophilous tundra herbs with non-specialized flowers. Among them are many species of saxifrages; only a few, e.g., *Saxifraga oppositifolia*, are obligately entomophilous. The small crucifers of the genus *Draba* have also an undoubted tendency toward self-pollination, and similar cases are known among the composites. However, many of the non-specialized herbs are truly entomophilous and are pollinated mainly by flies such as hover-flies, damsel flies, carrion flies and boll-weevils, etc. There is no basis for doubting the presence of entomophily among many of the rosaceous species or some of the ranunculaceous, caryophyllaceous and cruciferous ones.

Experiments effecting isolation of different plant species from insects during anthesis by means of netting bags have corroborated the assumption that many species of tundra herbs are actually able to manage without pollinators. Thus, when isolated, fully developed and fully viable seeds were formed by the saxifrages *Saxifraga punctata, S. nivalis, S. hieracifolia* as well as by the composites, Ilyin's arnica, dwarf fleabane and tansy (*Arnica iljinii, Erigeron eriocephalus, Pyrethrum bipinnatum*) and such species as cinquefoil (*Potentilla stipularis*), thrift (*Armeria arctica*), chick-weed (*Cerastium maximum*) and northern androsace (*Androsace septentrionalis*) and others.

However, some representatives of herbs do not set seed and are unable to reproduce without the aid of pollinators. This has been established for the Asiatic forget-me-not (*Eritrichum*) buttercup (*Ranunculus borealis*), spotted aven (*Dryas punctata*), and late flowering *Lloydia serotina*, etc. In a number of cases, seeds have been formed after isolation but in significantly lesser quantity than in the controls. This was demonstrated for Jacob's ladder (*Polemonium*) and some species of buttercup and saxifrage.

My own knowledge of the specialistic literature and my personal investigations have not convinced me that autogamy among tundra plants is more widespread than in associations within other zones. Recently I obtained some data regarding a significant occurrence of self-pollination among plants typical of the taiga zone. But the percentage of the typical entomophiles is not very high even in the forest steppe. The quantity and the variation of pollinators is at its greatest within the temperate zone. Evidently the presence of an alternative method of pollination (autogamy, anemophily, etc.) is a general characteristic of the majority of the entomophiles. Only a few species and small groups come out as winners in the competition for pollinators and have the capacity for retaining and perfecting obligatory entomophily.

Birds and invertebrate animals

Many birds are called entomophagous or insectivorous. This does not always mean that a particular species of bird feeding on insects will not consume other invertebrates as well, such as spiders, molluscs, millipedes etc. Actually anything under the convenient epithet of 'insect' is used (as long as we understand this is a conditional term). In the tundra zone both insects and other invertebrates are consumed by wading birds, passeriform birds, pochards, some gulls and, in individual cases, by predatory birds (for further details see Chernov, 1967).

Waders and snipe (the Charadrii) make up about one third of the arctic ornithofauna. Thus, in the Eurasiatic tundra there are about 30 species belonging to these groups out of a total of about 100 bird species. Practically all the tundra species are typically insectivorous. Only a few of them feed on plants as well. For instance the ruff (*Philomachus pugnax*) feeds to a great extent on the achenes of the buttercup *Ranunculus pallasii*, parts of viviparous plant species, etc. The stomach of the ruff is comparatively large and muscular and pebbles are always found in it (an indication that it is a herbivorous bird). The plovers often feed on green leaves, seeds and flowers.

There are, among the snipe and waders, species associated with lakes and mires as well as with 'dry land'; some inhabit the elevated moss tundras. Depending on the associations of these birds and their mode of feeding, the Charadrii in the tundra zone may be divided into a few groups. To the first of these groups (the Phalaropodidae) can be referred such interesting tundra birds as the phalaropes, adapted to feeding on water organisms which they mainly gather from the surface of the water. In the Eurasian tundra there are two such species, the red-necked or northern phalarope (*Phalaropus lobatus*) and the grey or red phalarope (*Ph. fulicarius*). The latter is absent from the continental tundra west of the Yenisey. These elegant birds swim well and while moving over the water with short strokes they scoop up small invertebrates from its surface. They often spin around in one spot while quickly scooping up the animals floating in the whirlpool thus created.

These waders usually search for food in small lakes and pools along beaches, adroitly dodging in and out among the aquatic plants. Their food consists not only of aquatic invertebrates like species of *Daphnia*, diving beetles, *Gammarus* and larvae of blood-sucking mosquitos, but also of terrestrial insects such as flies, mosquitoes, carabids and leaf-cutters, moths, etc. which are blown into the water by the winds, washed down from the moss sward along the banks by the waves and found, together with debris, in the surf belt. The waders love to feed in such places.

The phalaropes are not limited to the food they can find on the surface of the water but procure it also from the moss swards when they are flooded. They consume many of the larvae belonging to the genus *Prionocera* (a tipulid). In the stomach of such a wader up to 20 larvae of this type have been found (see Fig. 55). The larvae live in submerged and strongly water-soaked moss carpets in tundra mires and along lake banks. Around their hind parts they carry a halo of hairs on flat lobes, surrounding two openings (stigma). By means of these lobes the larvae adhere to the surface of the water film while the rest of the body is left hanging down among the mosses. Evidently the phalaropes are perfectly able to distinguish the larvae under such conditions, to which they may indeed be guided by the big and darkly coloured stigmata.

An abundance of chironomid larvae live in bodies of water in the tundra. The phalaropes gather small quantities from the ponds. After hatching from the pupa the adult chironomids come to the surface of the water where the phalaropes feed on them also.

   The phalaropes consume enormous quantities of microscopic invertebrates such as springtails, living on the surface of water films at the edge of ponds and lakes (the best known among them is *Podura aquatica*, but there are many other species as well.). They feed to a great extent on fresh-water crustaceans, *Daphnia* and *Cyclops*, but also on salt-water species such as copepods. Their stomachs can be completely stuffed with these minute animals alone with no other visible additions. Evidently, the phalaropes have very sharp vision and an astonishingly exact co-ordination of movement. One can only marvel at what they distinguish; they miss nothing and are able to locate these minute, hardly visible animals on the wave-tossed surface of the water among all the debris. The biology of the phalaropes is closely associated with the tundra landscape and the presence there of small bodies of water full of life, providing adequate food for these marvellous birds.

Fig. 55. Larvae of crane-flies, the basic food source of many insectivorous birds: 1, General view of larvae belonging to the genus *Tipula*; 2, pupa of a crane-fly; 3, hind part of a *Tipula* larva in the soil; 4, hind part of the aquatic larva belonging to *Prionocera*.

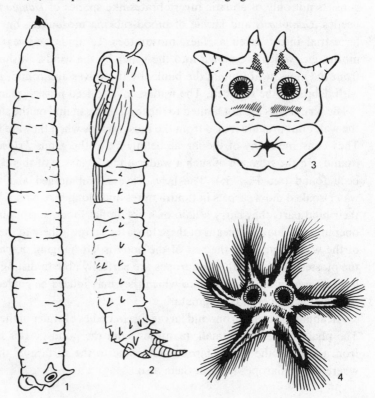

The second group comprises the species associated with mires: the little stint (*Calidris minuta*), the dunlin (*C. alpina*), the pectoral sandpiper (*C. melanotos*), the sharp-tailed sandpiper (*C. acuminata*) and the rufous-necked sandpiper (*C. ruficollis*) as well as the ruff (*Philomachus pugnax*) and some others, which spread from the south. In this group the tundra sandpipers predominate by their abundance; the little stint, the pectoral sandpiper and the ruff are especially numerous.

Species belonging to this ecological group are most typical of the two southern subzones. In the arctic tundra subzone some of them are lacking, others represented by small numbers only. The sandpipers nest in mires, marshy wetlands along the sea coasts and on low lands around lakes but often also on watersheds in the dry tundra. Excellent vision, hearing and perhaps some sense of smell assist them. By plunging their bills into mosses, sandpipers are indeed able to locate springtails and other squirming invertebrates. They also feed in shallow lakes and streams but rarely in the sea, although this does happen frequently toward the end of summer and in early autumn when they begin migrating.

Larvae of the long-legged genus *Prionocera* (a tipulid) predominate among the food species of these birds during the nesting period. These larvae make up no less than half of the bulk of their food. The sandpipers also consume the pupae of *Prionocera* and they fill their stomachs with adult mosquitos as well, on those days when these have hatched in great masses.

The tipulid larvae develop rather slowly: in the typical tundra subzone their development actually takes at least three years. In some years these mosquitos emerge in particularly large numbers. Another year there may be no mass appearance at all. This phenomenon is known also in other mosquito species with a development that extends over several years, e.g., the mayflies. The majority of these develop in synchrony and after a mass flight there follow years in the course of which a new population is being built up. In years with low numbers of *Prionocera* larvae, the number of nesting sandpipers may also be reduced. Caddis-flies and limonids (closely related to the tipulids), beetles and few-bristled aquatic worms, are among other food consumed by these birds but there is also some plant food, specially the achenes of the aquatic buttercup *Ranunculus pallasii*.

In this group may also be included Temminck's stint (*Calidris temminckii*), the broad-billed sandpiper (*Limicola falcinellus*) and occasionally also the pin-tailed snipe (*Gallinago stenura*), the jack-snipe (*Lymnocryptes minima*), the bar-tailed godwit (*Limosa lapponica*), the spotted redshank (*Tringa erythropus*) and some other species which extend into the tundra from the south. Each of these species has characteristic traits with respect to biology, feeding habits and biotopes. Data on their food are very scarce. Only a few interesting facts can be presented. Thus the pin-tailed snipe, known for their ability to gather food from wet soil, are true to this habit also in the tundra. In the stomachs of both the European snipe (*Gallinago gallinago*) and the Asiatic pin-tailed snipe (*Gallinago stenura*) there are usually only specimens of Nordenskiold's earthworm (*Eisenia nordenskioldi*), which is often consumed by tundra birds.

A third group is represented by terrestrial snipe and plovers (Charadriidae) colonizing the most elevated and driest habitats. The most typical are the plovers. Those favouring the driest sites are the black-bellied plover (*Pluvialis squatarola*) and the lesser golden plover (*P. dominica*). The black-bellied (or grey) plover is a circumpolar species, typical of the northern half of the tundra zone. The lesser golden plover is distributed within the typical tundra of both America and Siberia. On the European tundra they are replaced by the (European) golden plover (*P. apricaria*), which is, however, somewhat less arctic and does not reach the northern tundra subzone but is most common in the forest tundra. The fourth species, *Eudromias morinellus*, the dotterel, inhabits mountain ranges and arctic tundra.

To this group there also belong typical inhabitants of the polar desert and the arctic tundra subzone such as the curlew sandpiper (*Calidris testacea*), the knot (*C. canutus*), the sanderling (*C. alba*) and the purple sandpiper (*C. maritima*). When nesting, these sandpipers are the most typical inhabitants of the high arctic zonal types of landscape. Unlike mire species that belong to this group there is no clear-cut picture of their feeding habit. Their food seems to vary in composition from one season to another and also within the different subzones.

On the typical tundra where the insect fauna is rich enough, the food of those plovers preferring dry habitats is very variable. Basically it consists of adult insects, especially beetles, large Diptera, crane-flies, spiders, etc. In the stomach of the black-bellied plover (*P. squatarola*), the lesser golden plover (*P. dominica*) and the dotterel (*Eudromias morinellus*) it is possible to find samples of various beetles, some of them very rare, which can be located with difficulty only after a long search (e.g., beetles of the genus *Carabus*, large specimens of *Pterostichus*, leaf-cutters, the weevil *Lepyrus*, etc.). Also the dotterel often feeds in mires where it consumes larvae of *Prionocera* in large quantities. All the species belonging to this group consume more or less large numbers of earthworms as well.

The black-bellied (or grey) and the lesser golden plovers also catch many crane-fly larvae typical of dry soils and belonging to the genus *Tipula*, but not those of *Prionocera*. There is an especially close relationship between these larvae and the red-breasted sandpiper (*Calidris testacea*), one of the most characteristic species of the arctic tundra ornithofauna. In Taymyr the bulk of its nutrition consists of larvae of the crane-fly (*Tipula carinifrons*), a species very characteristic of this landscape. It lives in the surface layer of thinly covered or denuded soils and is easily dug out by the red-breasted sandpipers with their long, sensitive beaks. The larvae of this crane-fly are also consumed by the red knot (*Calidris canutus*), the sanderling (*C. alba*) and the black-bellied plover (*Pluvialis squatarola*). The purple sandpipers (*C. maritima*) inhabit the most northerly part of the tundra zone and parts of the polar desert where, locally, they are the only remaining representatives of the plovers. There they feed mainly on surface chironomids and their larvae living in the polar desert soil.

The content of the food of terrestrial plovers and snipe changes, thus, from being very various in the southern subzone to being very poor in the north, clearly reflecting the degree of variation among the invertebrate fauna. We should note still one more peculiarity in the feeding habits of these snipe: in the stomach of dry site species more so than in that of mire species many non-feed particles such as detritus, dead twigs, pebbles and such are found. This reminds us of the fact that the dry habitat snipe have less food available and are often starving.

A fourth group belonging to the so-called Mongolian plovers are inhabitants of sandy–pebbly river banks and maritime beaches. Most common in the tundra zone is the ringed plover (*Charadrius hiaticula*). It can be seen along river banks, the sea shore and also often on stony watersheds of the tundra. It is distributed far to the south, inhabiting also subtropical and tropical latitudes. On Chukotka there is another species of plover, the true Mongolian (*Ch. mongolus*). The original habitats of the ancestors of all the plovers were sea-shores or inland bodies of water with a firm bottom, but early on there was an adaptation in the food gathering habit, that is to search for food by visual means or to hunt for crawling or flying insects 'as these are escaping'. The ringed plover manages just in this manner. They feed on various insects living on the surface of sandy and stony beaches: flies developing in the debris of seaweed along the beaches, crane-flies, ground beetles of the genus *Elaphrus* and bugs belonging to the family Saldidae, etc.

They also consume large quantities of limnic invertebrates, both beetles and larvae of diving beetles, chironomids, etc., also fly larvae living in thick moss carpets where they cannot be located by eyesight alone. It is obvious that these plovers locate food also by 'feel' or by 'touch'. In ornithological literature a special adaptation of plovers for searching for food is described as 'reconnoitering the ground with their feet'. The disturbed larvae begin to squirm and thereby create a vibration in the soil or sand, making it possible for the plovers to locate them.

In more southerly landscapes the great majority of the birds belong to the Passeriformes. In the tundra zone passerines are fewer than those belonging to other orders. Thus, the tundra Charidrii amount to about 40% of all species belonging to that group in the USSR, but Passeriformes represent only 6% (or 22 species). Such a strong impoverishment of this order in the tundra is first and foremost linked to the food situation. The passerines are less adapted to finding food in soil or below the moss than other birds; they feed mainly on what can be located in the open and on the surface. They are, of course, not adapted to gathering food from bodies of water, the richest food source in the tundra zone biotope to which the zoophagous tundra birds such as plovers, snipe, diving duck and so on are linked. There are no species of passerines adapted to feeding on coarse plants such as are eaten by ptarmigan and willow grouse. Only a few species of passerines cover the zonal tundra landscapes to the same extent as many of the plovers. There are not more than five such species or, if taking a more restricted point of view of this problem, actually only three.

The majority of the passerines feed to a greater or lesser extent on insects and other invertebrates. The nestlings are fed only animal food. It is possible to distinguish three rather conventional groups: mainly herbivorous species (feeding on seeds or grain); species preferring a mixed diet; and insectivorous species. Exclusively herbivorous small birds such as the crossbill (*Loxia curvirostra*) are not found in the tundra zone. To the mainly seed-eating birds belong the redpoll and the arctic redpoll (*Acanthis flammea, A. hornemanni*) which are common in the southern shrub–tundra but are also found locally on the typical tundra.

Within the group preferring a mixed diet, three species are widely distributed in the northern subzone: the Lapland bunting (*Calcarius*

Fig. 56. Typical small tundra birds with variable food habits. 1, Snow bunting (*Plectrophenax nivalis*); 2, Lapland bunting (*Calcarius lapponicus*); 3, shore lark (*Eremophila alpestris*).

1   2   3

lapponicus), the shore lark (*Eremophila alpestris*) and the snow bunting (*Plectrophenax nivalis*); see Fig. 56. They can be considered the most characteristic of the tundra passerines. For these species a rather significant variety in the character of the food is typical. Seeds of various plants are the basic source of food, but insects constitute an essential addition to their diet. At the time when the insects are most active, these birds may feed as intensely on them as true insectivores do. In years with a late spring and a cold summer adult birds use mainly seeds for food, but the nestlings are always fed insects only. The Lapland bunting is a species feeding mostly on plants. These birds are not able to pick insects from the moss carpet but can gather them more easily from the soil; their prey is mainly active or flying forms of weevils, small beetles, to a lesser extent caddis-flies, leaf miners and various kinds of flies, etc.

To the group preferring a mixed diet belong also the little bunting (*Emberiza pusilla*), Pallas' reed bunting (*E. pallasi*), Naumann's thrush (*Turdus naumanni*) and some others reaching as far as the southern part of the subarctic.

Among the typically insectivorous small birds within the tundra zone we find the red-throated pipit (*Anthus cervinus*), the white, the yellow (or blue-headed) and the citrine wagtails (*Motacilla alba*, *M. flava*, *M. citreola*), the bluethroat (*Luscinia svecica*), the wheatear (*Oenanthe oenanthe*), the arctic warbler (*Phylloscopus borealis*), the chiffchaff (*Ph. collybita*), and a few other species.

The red-throated pipit (*Anthus cervinus*) stands out sharply as being the only one of this group inhabiting landscapes in great numbers around watersheds. However, it clearly gravitates towards the southern belt of the subarctic. Its highest concentration occurs on the southern tundra and on the forest tundra. The food of the red-throated pipit varies widely. It consists mainly of insects which it picks out of moss swards and leaf litter under shrubs as well as from grassy sward. It consumes notably great quantities of caterpillars, larvae of sawflies, leaf beetles living on the willow leaves and herbaceous plants, but also small animals such as mites.

The bluethroat (*Luscinia svecica*) is met with where there are small stands of willows. Locally it reaches north above the limit of the willow shrubs and then nests in grass tussocks or on stony slopes. It often behaves like a true synanthrope (i.e., a species living among people) and builds its nests among boards, wood shavings, rolls of wire, etc. All this characterizes it from an ecological point of view as an extraordinarily plastic species. Where there are bushes the bluethroat feeds mainly on the different inhabitants among willows such as litter forms (especially of carabids) and insects living in the upper layers. The caterpillars of tortricid moths are very often found in its stomach together with catkin weevils, sawflies, leaf suckers (*Psylla*), etc. The young individuals visit human settlements at the end of the summer where they feed on larvae crawling and pupating, or hatching, in refuse, such as those of carrion flies.

The wheatear (*Oenanthe oenanthe*) is distributed throughout all the zones. Its nesting sites are ravines and stony slopes, pebbly spits overgrown with wetland vegetation, overhanging river banks and crumbling slopes. The food of the wheatear is limited to slow-moving surface forms; like all the tundra insectivores it consumes large quantities of crane-flies when these occur in great numbers during summer. Beetles play an important role in its diet, especially large species such as carabids (*Pterostichus vermiculosus*), carrion beetles (*Thanatophilus lapponicus*), weevils (*Lepyrus arcticus*) and water beetles (*Colymbetes* and *Graphoderus*).

A characteristic trait in the food habits of the wheatear is the consumption of large numbers of bumble-bees, which are rarely touched by other birds. In the stomach of the wheatear can be found species such as *Bombus hyperboreus*, *B. lapponicus* and *B. balteatus*. Its nest sites coincide with the main habitats of these bumble-bees, i.e., herbaceous associations along banks or shores, steep slopes, entomophilous herbage, etc. The wheatear has, on the whole, a tendency to feed mostly on the larger species of insects. The well-known inclination of that species to synanthropism is even more apparent at high latitudes. While in southern latitudes this behaviour is apparently conditioned by nesting opportunities, it is no less important with respect to its feeding habits. When nesting in human settlements the wheatear feeds on larvae and imagos of carrion flies and carrion beetles.

The white wagtail (*Motacilla alba*) is also widely distributed within the tundra zone. Here, as throughout its extensive range, it is associated mainly with coastal areas and human habitation.

Other typical insectivorous small birds occur only in the southern parts of the tundra zone where the conditions for insectivory are always favourable. Here the largest concentrations occur of bog species, the yellow wagtail (*Motacilla flava*) and the willow warbler (*Phylloscopus trochilus*) which are associated with shrubs.

However, in the northern subzone, species of small birds predominate which feed on a mixed diet but they are also able at any time of the year to subsist solely on plant food. To these belong the snow bunting (*Plectrophenax nivalis*), the Lapland bunting (*Calcarius lapponicus*) and the shore lark (*Eremophila alpestris*). Only about ten species feeding on predominantly animal food are recorded in the subarctic. The majority of these are limited in their distribution to the southern tundras. Such species as the bluethroat (*Luscinia svecica*), the wheatear (*Oenanthe oenanthe*) and the white wagtail (*Motacilla alba*) occupy ecological niches within the intrazonal parts of the landscape. The red-throated pipit (*Anthus cervinus*) is the only species of this group which is widely distributed in zonal associations but also prefers the southern belt of the subarctic.

From the point of view of feeding the shrub belt of the southern tundra offers more favourable conditions for insectivores: there is on the one hand a complex of invertebrates in the leaf litter, and on the other an abundance of species living in the open within the shrub tier (e.g., aphids, leaf suckers, larvae of sawflies, tortricid moths and a lot of early spring forms associated with catkins, hover-flies, bumble-bees, etc.).

Nine species belonging to the order of gulls (Laridae) nest in the tundra zone. Among them there are three species of skua, the pomarine skua (*Stercorarius pomarinus*), the arctic skua (*S. parasiticus*) and the long-tailed skua (*S. longicaudus*), the herring gull (*Larus argentatus*) as well as the arctic tern (*Sterna paradisaea*). The glaucous gull (the so-called 'burgomaster', *L. hyperboreus*) prefers the high-arctic areas. The common gull (*Larus canus*) penetrates into the tundra zone as far as Poluostrov Kanina. In different areas of arctic Siberia Ross's gull (*Rhodostethia rosea*) and Sabine's gull (*Xema sabinei*) are distributed.

The skuas are predominantly predatory, feeding on rodents, birds and fish. Insects play a secondary role in their diet. The smallest of them, the pomarine skua, consumes large insects. The importance of insects in the diet of the small sized long-tailed skua is sometimes fairly great. It consumes large quantities of crane-flies when they occur in masses but also various easily seen, slow-moving beetles, such as carabids, leaf beetles and carrion beetles. Insectivory among these species of *Stercorarius* appears especially distinct during years when the number of lemmings is low. It is not known whether the large gulls – the herring gull and the glaucous gull (*Larus argentatus*, *L. hyperboreus*) – ever eat insects to any major extent although it is possible to find large quantities of carrion beetles in their stomachs. These may be gulped down together with the bodies of dead lemmings.

The food habits and the general ecology of two interesting species of gulls, Sabine's gull (*Xema sabini*) and Ross's gull (*Rhodostethia rosea*), are not well known. These elegant small seabirds, with very beautiful plumage, nest on the tundra but winter further north, migrating during the polar night to the open waters in the Arctic Ocean. During nesting time they are typically insectivorous. Their food includes terrestrial insects but especially adult crane-flies, consumed in large quantities during the imago stage, together with various bog and aquatic invertebrates, consisting mainly of fairy shrimps, molluscs and amphipods, but also of chironomids and caddis flies.

The arctic tern (*Sterna paradisaea*) is a narrowly specialized bird. Its food is limited to specific food biotopes, which make use of water surfaces. Besides truly aquatic species, an essential role is played by terrestrial insects which are blown out over water by the wind. Chironomid chrysales are of great importance and are consumed during the time when they occur in large numbers on the surface of water following the imago stage. About half of the food consumed by these terns consists of fish.

When comparing the food habits of the different species of gulls on the tundra one finds an inverse relationship between the size of the bird and its level of insect consumption. Thus, among the skuas the essential role of insects in the food is associated with the smallest of the species, the long-tailed skua (*Stercorarius longicaudus*). The smallest gulls – Sabine's (*Xema sabinei*) and Ross's gull (*Rhodostethia rosea*) – as well as the arctic tern (*Sterna paradisaea*) are predominantly insectivorous during the nesting time.

An important place among the tundra insectivores is occupied by the ducks: the long-tailed duck or oldsquaw (*Clangula hyemalis*), the scaup (*Aythya marila*), Steller's eider (*Somatoria stelleri*), the king eider (*S. spectabilis*) and the spectacled eider (*S. fischeri*) and some others. They fly in from the sea to their nesting sites and switch to feeding on fresh-water invertebrates located by diving to the bottom of tundra ponds. On the Yugorskiy Poluostrov the larvae of chironomids have been found in the stomach of the long-tailed duck by the 25th May but in the eastern sector of the subarctic, of course, somewhat later.

Everywhere during summer the main stomach content of the diving ducks (the Aythyinae) consists of chironomid larvae. Up to 2000 specimens or about 22 grammes have been found in the stomachs of eiders. An essential role is played also by the limnic larvae of crane-flies (Tipulidae). There were 30 specimens in the stomach of a long-tailed duck. These birds consume also large quantities of caddis-fly larvae (up to 100 specimens have been found in the stomach of a king eider). During summer another food category consists of diving beetles, fly larvae and shrimps. The diving ducks also consume plant food such as the achenes of the aquatic buttercup *Ranunculus pallasii*. Vegetative parts of plant are usually mixed in and swallowed when they gather benthic material.

It is possible to distinguish some groups of insects filling an especially important role in the diet of insectivorous birds. Primarily these are the chironomids and the crane-flies or tipulids. The former represent the basic food of all vertebrates associated with tundra ponds. Evidently the nesting density of the diving ducks (as well as the stock of fresh-water white fish) is determined mainly by the abundance of the chironomids. They are used by wading birds and snipe as supplementary food; during the chrysalis stage quantities of them are consumed by the arctic tern. The adult chironomids are devoured by all the terrestrial birds.

A major role in the trophic association of tundra birds is played by the crane-flies (Tipulidae). Many birds consume an equal amount of both larvae and adults. The semi-aquatic larvae form the main food basis of many snipe. In some areas the abundance of these larvae actually determines the nesting density of the snipe on the mires. At the time of mass flight, which occurs within a short span of time and is very intense, the adult crane flies are also being eaten by all kinds of Charadrii, passerines and gulls. These tipulids possess a whole number of characteristics making them trophically important for insectivorous birds. First of all we should mention their large size, great numbers, habitation in the tundra and slow movement. The last-mentioned trait is very typical of crane-flies which usually creep over the ground or fly slowly for a short distance only. It is typical of arctic species occurring in great numbers, that they have reduced wings and, thus, a diminished capacity for flight.

Under conditions in the tundra zone and in spite of the great stress on the trophic relationships, a major part of an important group of animals is not included in the diet of the birds and are not used in proportion to their numbers. Thus, soil invertebrates are not often consumed. Soil larvae and earthworms (*Eisenia*) are regularly found in the stomachs of the curlew sandpiper (*Calidris testacea*), the black-bellied plover (*Pluvialis squatarola*) and the lesser golden plover (*P. dominica*), but also in those of some snipe (*Gallinago* spp.) extending further south.

The birds of the mires consume few of the few-bristled worms inhabiting submerged mossy sward. These mobile worms crawl among entangling mosses and break up easily when attempts are made to pull them out. Because of this ability they do not often end up in the stomachs of the snipe feeding at such sites.

The ecological niche represented by birds hunting on the wing is not well known either. The nesting area of such groups of birds is limited to the forest tundra in the northlands. In the neighbourhood of Noril'sk the house martin (*Delichon urbica*) and the sand martin (*Riparia riparia*), which is also called the bank swallow, are still common. I have myself observed swallows (*Hirundo rustica*) and swifts (*Apus apus*) flying over the Yugorskiy Poluostrov. Penetration of these insectivores further northwards is evidently hampered by strong fluctuations in the activity of insects there and the very short period during summer when these occur in large quantities.

Examples of very definite food specialization among birds are found more or less associated with an aqueous environment and adaptation to feeding on aquatic organisms. Thus, within the group of gulls, the most narrowly specialized is the arctic tern (*Sterna paradisaea*). Among waders and snipe, the most narrowly specialized are species of phalaropes. For terrestrial species, e.g., the plovers, a high degree of polytrophy is typical. The most numerous of the small birds, the snow bunting (*Plectrophenax nivalis*), the Lapland bunting (*Calcarius lapponicus*) and the shore lark (*Eremophila alpestris*) are euryphages. Other species of small birds are only generally insectivorous.

Aquatic invertebrates included in the diet of birds are far more plentiful than the terrestrial species: this can be explained by their accessibility in bodies of water in the tundra. The strong periodicity of their maximum development by comparison with that of terrestrial species is a most essential characteristic of birds. This is related to the long period when the ice is melting and the long time the bodies of water stay cool in contrast to the surface layers of the soil. At the end of the arctic summer, in August, terrestrial insects stop developing, disappear, go into hibernation or live hidden deep within the soil or the moss carpet. At this time aquatic organisms are at their climax: small crustaceans and larvae of chironomids multiply in enormous quantities. At the end of the summer mass development of marine plankton, especially of crustaceans, also takes place.

When the new generation of young birds takes to the wing, the majority of waders and snipe have switched to feeding on aquatic invertebrates. Birds not adapted to gather food from the water begin to eat plant food at the end of the nesting period. It should be noticed that only three or four species of birds feed exclusively on terrestrial insects throughout the course of their soujourn in the tundra zone. Of these only the red-throated pipit (*Anthus cervicus*) is comparatively widely distributed over the tundra landscape. The others are found locally and mainly in intrazonal habitats within the landscape.

Birds of prey

I am obliged mainly to Osmolovskaya, who in 1948 published her work: 'Ecology of the Predatory Birds of Yamal', for the information on the food relationships of tundra birds of prey. During the period when field work for the present book was carried out, I was able to convince myself of the accuracy with respect to the basic conditions as reported by Osmolovskaya. They were also corroborated during an analysis of the food relationships between birds and insects (Chernov, 1967).

In the Eurasian tundra some ten species of birds are met with which can be referred to as predatory. There are six species of daytime predators among which are the rough-legged buzzard (*Buteo lagopus*), the white-tailed or sea eagle (*Haliaeetus albicilla*) and the hen-harrier (*Circus cyaneus*). To the falcons belong the peregrine (*Falco peregrinus*) and the gyrfalcon (*F. rusticolus*) as well as the merlin (*F. columbarius*). There are two species of owl, the snowy owl (*Nyctea scandiaca*) and the short-eared (*Asio flammeus*). Here belong also three species of skua, the long-tailed (*Stercorarius longicaudus*), the arctic (*S. parasiticus*) and the pomarine skua (*S. pomarinus*).

The hawks, the falcons and the owls of the tundra almost exclusively consume warm-blooded animals, that is, birds and rodents. The skuas, taxonomically closely related to the gulls, feed on a mixed diet including even plant material; they are, in other words, omnivorous or polytrophic, that is, adapted to eating a very variable diet. However, predation predominates for all these species.

In addition to the species enumerated we also find the raven (*Corvus corax*) here, which is also known to be a predator. The epithet 'predator' can be applied also to some fish-eating birds such as gulls and loons but this concerns aquatic life, and so is not treated in this book.

Of the species mentioned above not all play an essential role within the biocoenoses of this area. The gyrfalcon (*Falco rusticolus*) is often erroneously included among the typical inhabitants of the tundra, but this large falcon is distributed mainly within the forest–tundra, the northern taiga, along the coast and in the mountains of the subarctic area. At the present time the gyrfalcon is on the verge of being exterminated by man and has become rare everywhere. The sea eagle (*Haliaeetus albicilla*), widely spread over many zones, the hen-harrier (*Circus cyaneus*) and the merlin (*Falco columbarius*) are regularly met with in the southern parts of the subarctic. The sea eagle often roams the tundra but rarely nests there. The merlin is common in the forest tundra and only rarely flies further north. This is a typical forest bird and it nests in the subarctic only where there are some trees or tall bushes. The short-eared owl (*Asio flammeus*), also called the bog-owl, nests regularly in the southern subzone in large numbers in years when rodents are abundant. It is rare in the typical tundra.

Only six of the predatory birds are always to be found and are common inhabitants of the tundra, exerting a considerable effect on life within its biocenoses. These are the snowy owl (*Nyctea scandiaca*), the rough-legged buzzard (*Buteo lagopus*), the peregrine falcon (*Falco peregrinus*) and the three species of skua (*Stercorarius longicaudus*, *S. parasiticus* and *S. pomarinus*). Among them the snowy owl is the most typical inhabitant of high latitudes. Its main nesting area is the arctic tundra subzone. The rough-legged buzzard is a species with quite a wide ecological amplitude. It nests all over the subarctic but reaches its greatest density within the typical tundra.

The peregrine falcon is a cosmopolitan species, but nowhere do its numbers reach as high a level as in the subarctic. It is evident that the tundra is the most favourable landscape for this species; there it finds open space, an abundance of birds, and lack of competitors as well as nesting sites available on precipices, ledges, slopes, etc. There is, however, no direct connection between the origin of this bird and the arctic landscapes.

All three skuas are typically omnivorous and distributed over all the zones of the tundra territory. The most common of them is the long-tailed skua (*Stercorarius longicaudus*). The pomarine skua (*S. pomarinus*) is more inclined toward high latitudes. Most of them prefer to nest within the arctic tundra subzone.

The prey of feathered predators consists of birds and small animals, mainly lemmings, but in the southern zone also of field mice. There are no frogs, lizards, snakes or large insects such as are the main supplements of the basic diet of hawks and falcons at more southerly latitudes. The predators mentioned have different hunting abilities which leads to their specialization, behaviour and numbers of their prey. Osmolovskaya distinguished four basic modes of hunting behaviour:

1. Swooping down on or chasing prey: typical mainly of falcons. They overtake small birds in the air, boldly attacking them at an angle from above and catching them with their claws. The peregrine falcon (*Falco peregrinus*) hunts in just this manner. This can be called the most perfect type of hunting used by birds but it makes the hunter dependent on the intended prey: it can be captured only in open areas or when in the air.

2. Lying in wait or ground-based surveillance: this is more typical of owls (*Nyctea, Asio*) which can sit for an extended period of time on a perch watching out for lemmings. Skuas (*Stercorarius*), too, often use this method and sometimes also the rough-legged buzzard (*Buteo lagopus*) does the same.

3. Aerial surveillance: this is the usual mode of the rough-legged buzzard. It soars slowly above the ground and sometimes it comes almost to a stand-still while hovering and beating its wings in the same way as the kestrel (*Falco tinnunculus*). The skuas, too, occasionally behave in this manner.

4. Stalking the prey: that is, flying low at a height of 3–5 m and suddenly swooping down on the surprised prey, scattering them in all directions. This is typical of the short-eared owl (*Asio flammeus*). The skuas also do this and sometimes the rough-legged buzzard and the snowy owl as well.

Thus, each species has its preferred method of hunting which it has perfected. It is, however, very typical of tundra predators to combine methods in order to utilize fully the basic food sources offered to them on the tundra. The skuas are especially successful. They are very agile,

Fig. 57. Skuas – the 'true pirates of the tundra' (photo.:
A. V. Krechmar); above: Arctic skua (*Stercorarius parasiticus*); below:
long-tailed skua (*S. longicaudus*).

constantly 'poking about' in the tundra and finding prey (see Fig. 57). Their favourite food is the lemming. They often attack them or birds, ravaging nests and eating eggs and baby birds. The well-known investigator of animals on the tundra, Sdobnikov (1953), called them 'aerial pirates'. The rough-legged buzzard hunts in a fairly varied fashion. In addition to its basic food of lemmings, it hunts also for stoats (*Mustela erminea*), young polar hares (*Lepus timidus*), and various kinds of birds from waders and snipe to small birds. The snowy owl is more conservative. During its nesting period it subsists almost exclusively on lemmings and takes only a very small number of birds (see Fig. 58).

The food of the peregrine falcon (*Falco peregrinus*; Fig. 59) consists mainly of ptarmigan and willow grouse (*Lagopus mutus*, *L. l. lagopus*) but also of pochards and dabbling ducks (Aythyinae, Anatinae), waders and small birds. Of course birds of medium size predominate, especially the ruff (*Philomachus pugnax*), the golden plovers (*Pluvialis dominica*, *P. apricaria*), the pin-tailed snipe (*Gallinago stenura*), the spotted redshank (*Tringa erythropus*) and the bar-tailed godwit (*Limosa lapponica*). The peregrine is, however, not limited in its hunting methods and often takes food on the ground to which the many remnants of lemmings in its nest bear witness.

The mode of hunting becomes less characteristic when there is a shortage of food. Thus, during years with low numbers of lemmings the rough-legged

Fig. 58. The snowy owl (*Nyctea scandiaca*), which feeds almost exclusively on the lemmings on the tundra (photo.: A. V. Krechmar).

buzzard (*Buteo lagopus*) hunts more for birds and the peregrine falcon (*Falco peregrinus*) will, during prolonged periods of bad weather or strong winds, hunt for lemmings more frequently. The dimensions of the hunting area also changes, depending on the availability of the basic food. When the lemmings are abundant, arctic owls and rough-legged buzzards hunt close to their nests, often within a radius of 2–3 km; when rodents are scarce they must fly over much longer distances. The size of the hunting territory of the peregrine falcon depends on the structure of bird colonies. In areas with an abundance of birds the hunting territory is small. Often this falcon returns to its nest with prey within three to five minutes of flying off, but when the falcon hunts for ptarmigan and ducks it may have to fly distances of 10 km or more.

The birds on which the predators feed vary within the different subzones. This is first and foremost related to the species composition of the likely prey. In the southern tundra subzone there are in addition to lemmings some species of vole, the narrow-skulled and Middendorff's (*Microtus gregalis, M. middendorffii*) and locally the root vole (*M. oeconomus*) as well as the large water vole (*Arvicola terrestris*). These rodents are consumed by rough-legged buzzards, short-eared owls and skuas to a greater or lesser extent. However, the main food of these birds consists of those animals which are the most abundant, that is, lemmings. Of the rodents, only lemmings occur in the northern half of the tundra zone. The peregrine

Fig. 59. The peregrine falcon (*Falco peregrinus*), the principal predator of medium sized birds (photo.: A. V. Krechmar).

falcon finds more favourable feeding conditions in the southern half of the tundra zone where there are many ptarmigan and snipe of medium size. Further north it more often catches diving sea ducks (scoters), especially the long-tailed duck (*Clangula hyemalis*), but also small birds and then mainly the Lapland bunting (*Calcarius lapponicus*).

Great numbers of bird species or rodents ordinarily predominate within the area of predators. However, in a number of cases a selection is possible, but the prey taken from a particular species may not be proportional to its distribution in the landscape. This may differ according to the so-called 'active' or 'passive' selection. Passive selection is determined by the behaviour of the prey itself. When perusing different literature sources regarding the diet of tundra predators, it is easy to find that the Lapland bunting (*Calcarius lapponicus*) is almost always listed among the small birds taken as prey. According to the data of Osmolovskaya, the merlin (*Falco columbarius*) hunts these buntings almost twice as often as all other birds taken together. Basically this species is met with also as a food source for the rough-legged buzzard (*Buteo lagopus*) as well as the snowy owl (*Nyctea scandiaca*). It is also to some extent hunted by the peregrine falcon (*F. peregrinus*).

The Lapland bunting is a flocking species. It is, however, not only a question of numbers. This is an inhabitant of open spaces, a bird with a relatively weak flying ability. During the nesting period the males thresh about and often sing when flying high into the air, then slowly lower themselves to the ground during the mating display. During such moments it becomes easy prey to falcons. At the end of the nesting period the adult buntings moult intensely and fly very badly. During that period they are often hunted by skuas. Other passerines, e.g., the red-throated pipit (*Anthus cervinus*), the shore lark (*Eremophila alpestris*) and the snow bunting (*Plectrophenax nivalis*) are, in spite of their large numbers, known less frequently to become prey to predators. The red-throated pipit and the shore lark are very active and the snow bunting nests in habitats which are well concealed (on rocky precipices, slopes, river banks, etc.) and they are able to hide themselves in a moment.

Passive selectivity occurs also in the feeding behaviour of the rough-legged buzzard (*Buteo lagopus*). In the southern tundra, Middendorff's vole and the narrow-skulled voles (*Microtus middendorffii, M. gregalis*) are locally more numerous than the lemmings. Among the prey within the territory of the rough-legged buzzard the lemmings predominate. By comparison, the voles lead a more hidden kind of life, but are always very active. The narrow-skulled vole is taken less often; it lives in colonies and builds a system of connecting burrows into which it can escape quickly when necessary.

The lemmings themselves also differ in the numbers taken by predators. The collared lemming (*Dicrostonyx torquatus*) is the less active, therefore it is more intensely preyed upon than the Siberian lemming (*Lemmus sibiricus*). The collared lemming predominates as food for rough-legged buzzards and snowy owls. The Siberian lemming is very active; within its colonies it constructs a multitude of tunnels where it can quickly elude its enemies. According to observations made by Osmolovskaya, the owls and the rough-legged buzzards hunt more often for the Siberian lemming during the periods when this is feeding and is easily caught by surprise, but they take the collared lemmings during moments when they are running around over the tundra.

Concerning the food habits of animals, protective colouring is of great importance together with the shape and size of the prey. These characteristics have developed due to the process of natural selection and the effect of predation. These traits are more distinct in the case of the collared lemming, which is subjected to more intense predation. During summer, its variegated yellow–brown–grey colouring stands out among the lichens and mosses of the tundra. In winter its white fur acts as camouflage against the snow.

Protective colouring does not, however, offer complete safety from enemies because during the process of evolution the capacity of hunters to discover their prey has also been perfected. There are, however, examples of very effective and functional protective adaptations. Osmolovskaya mentions interesting details from the feeding behaviour of predators. In the course of, at the most, two weeks there are in the tundra very many downy fledglings of Charadrii, which because of their conduct and size are, so to speak, prone to be preyed upon by birds hunting for lemmings. At the same time, however, they practically never fall victim to feathered predators. This may actually be the result of their perfect, protective coloration fully matching that of the moss tundra and also of their ability suddenly to become absolutely still when danger threatens.

Due to the process of natural selection young Charadrii have been excluded from the range of food of predatory birds. The latter have experienced no counter-adaptation due to the fact that their hunting adaptation is directed towards preying upon mouse-like rodents. If the downy chicks had not such a natural capacity for protection, many predators of the tundra using them as a supplementary source of food might already have exterminated the major part of the populations of Charadrii because the nesting density of these birds is extremely low in comparison with the enormous numbers of lemmings available.

The above examples illustrate passive selectivity, the most widely distributed type in the tundra. Active selectivity is the ability of predators to select among the available food and take that of most value to them. For instance, many insectivores actively search out larvae with fleshy integuments and especially caterpillars but do not touch beetles or ants. The development of such an ability is possible in cases where many kinds of food are generally available. In the subarctic active selectivity is restricted by the low diversity of food objects. For their diet the carnivorous animals in the tundra usually track down for preference the predominant and most available kinds of prey. In part this is characteristic also of such specialized predators as the peregrine falcons. They hunt for birds in the tundra while clearly following the principle of 'first come, first caught'. Above, it has been demonstrated that insectivorous birds on the tundra behave in the same manner.

No doubt active selectivity has also developed to a certain extent among the tundra predators but there are hardly any definite data regarding it. Probably the predominance of medium sized birds in the diet of peregrine falcons is what is most advantageous from the point of view of energy expenditure. These falcons cannot in general take big birds such as ducks on the nest and devour them then and there. Small birds provide, however, only small quantities of food but their capture and transportation to the nest requires still no less time and energy.

Concerning the composition of the basic diet, the tundra predators can be distinguished into two groups, the ornithovores (those preying on birds) and rodentivores (those preying on rodents). The falcons belong to the former; to the latter belong owls, rough-legged buzzards and skuas. These groups differ in respect to their ecology and population dynamics. The nesting density of herbivorous and insectivorous birds preyed upon by falcons varies little over the years, as they in turn depend on their own source of basic food. The fluctuation in numbers of the different species is not strongly affected by the fact that they are fed upon by the falcons, because the amount of preying hardly ever varies. The number of ornithovores is also fairly constant in the course of time.

The observed decrease in numbers of peregrine falcons and gyrfalcons (*Falco peregrinus, F. rusticolus*) in recent times can be blamed to a great extent on anthropogenic causes, that is, the direct extermination by man of these birds or his depriving them of their basic food, for instance by hunting willow grouse (*Lagopus l. lagopus*), their main food in the southern taiga. Except for interference by man, the density of peregrines in the subarctic would be exceptionally stable. From year to year they occupy identical nesting sites, distributed at a definite distance from each other. This has been mentioned by Zhitkov (1913) and Osmolovskaya (1948) from Yamal, and by Krechmar (1966) from Taymyr. In the presence of suitable sites (river banks, slopes), nests of peregrines in the southern tundra subzone are distributed at a distance of *c.* 8 km from each other and in the typical tundra zone about 15–20 km apart. These birds are, however, hardly ever seen nesting within the enormous areas of the gently undulating watersheds.

Counts that have been made indicate that at such a nesting density the falcons will remove only a small portion of the tundra populations of small birds, that is, not more than 5%. It is thus quite evident that the association of predatory birds is rather weak and will vary in relation to habitats, composition of the bird community, the season, etc. At the same time it is obvious that the stable and generally not very high degree of nesting density among falcons in the tundra has been assured by selection of an optimal variant of a system of 'predator/prey relationship' under the circumstances given. No doubt falcons can have only a positive effect on the lives of their prey species by removing preferably the sick and weak individuals.

A different aspect is shown by the interrelationship between predatory birds and rodents. The collared lemming (*Dicrostonyx torquatus*) and the Siberian lemming (*Lemmus sibiricus*) are the main food sources of the majority of tundra predators (both feathered and furry). Snowy owls (*Nyctea scandiaca*), rough-legged buzzards (*Buteo lagopus*) and skuas (*Stercorarius* spp.) as well as stoats (*Mustela erminea*) and foxes (*Alopex lagopus*) live basically at the expense of these rodents. The role of lemmings is the most important during the period when the nestlings are reared. The adult birds are still able to vary their food, consuming both birds and insects, and in the southern subzone other rodents and even berries (as does the long-tailed skua, for example). But they rear their young mainly on lemmings, something that is especially typical of snowy owls and rough-legged buzzards.

However, the number of lemmings is not stable; it is subject to sharp variation from year to year. An answer to the question 'why?' has still not been elucidated in spite of much literature devoted to it. Some authors suggest a cyclicity in the numbers of lemmings and give it a 3–4 year periodicity. Others construct data disproving any definite cyclicity. Only one thing is sure, the number of lemmings varies widely from time to time. During years of maximal occurrence lemmings are met with in enormous numbers (they propagate mainly under the snow during winter) and they feed extensively on the vegetation. During such times their behaviour also changes. They become less cautious, often very aggressive, do not hide and will, squeaking, even attack an approaching human. At other times they become sluggish and inactive. This is indeed an example of an epizootic, occurring among overcrowded populations. During such a period lemmings are devoured in great numbers by all the tundra predators, even by reindeer (*Rangifer tarandus*). During spring, many lemmings fall into lakes and rivers where they are eaten by diving ducks such as eiders but also by predatory fishes such as the salmon (*Salvelinus alpinus*). I have caught large specimens of salmon with their stomachs literally stuffed full of lemmings (about ten carcasses in one stomach!).

After attaining their maximum number, the lemmings begin to decline rapidly. The causes for this are rather variable. On the one hand their fertility declines while, on the other hand, disease becomes rampant among the animals. Recently both tularaemia and leptospirosis have been confirmed among lemmings and other voles in both Yamal and Taymyr. It stands to reason that an epizootic alone cannot have such a decisive effect on the numbers of these animals. According to Dunayeva (1978) at most only 7% of the dead lemmings had succumbed to tularaemia. Evidently many other factors are operating in addition to disease: predation, food shortage, lowered fecundity of the females and of the quality of the animals mating (a consequence of overcrowding within a population) and death from exposure during the time of very active roaming around over the tundra, etc.

Following the decline of the lemmings a large number of carcasses remain on the tundra. These are consumed by fly larvae of which the mass-occurring arctic species of bluebottle or the carrion flies are especially typical, e.g., *Boreellus atriceps*. A moss, *Tetraplodon mnioides*, begins to grow on the remains of their bones. Their bright green, roundish cushions are sprinkled over the tundra. Also species of mites and nematodes specialize in feeding on carcasses.

The food base of the rodentivores is, thus, very unstable, which in turn determines the strong fluctuation in nesting density and fertility of snowy owls, rough-legged buzzards and skuas from year to year. The characteristics of bird associations also vary. In years with maximum numbers of lemmings rough-legged buzzards are distributed at a distance of 2–3 km from each other and have hunting territories on an average covering 7 km². Often one can find the nests of snowy owls near by on stony ridges, and at sites on level tundra there are the nests of the skuas. Osmolovskaya estimated that one family of rough-legged buzzards needs about 2000 lemmings during a summer season. At a density of 20–40 such animals per hectare, this amounts to about 10% of the lemming population. All the predators together may, indeed, exterminate up to 30% of these rodents. This is a rather impressive figure. At such a rate of preying on the stock of lemmings, the birds can play a decisive role in the reduction of their numbers.

During years with large numbers of lemmings on the tundra, the numbers of predatory birds mating is overwhelming. Therefore they lay many eggs, considerably more than their southern relatives. Thus, during years of lemming abundance, the rough-legged buzzard often lays five eggs and sometimes as many as seven; the snowy owl often lays from five to nine, but sometimes as many as 11. Rough-legged buzzards, ordinarily nesting in the temperate belt, have clutches consisting of two to four eggs as a rule. The majority of forest owls lay about five eggs. During years with low numbers of lemmings, the density of nesting predators is reduced. Many adult birds do not nest at all. The eggs in their clutches are fewer; the rough-legged buzzard lays two to three, the snowy owl up to three. The skuas almost always have two eggs to a clutch and they react to the decline in numbers of lemmings mainly by a reduction of nesting pairs.

In the literature little is said about the fertility of the tundra birds, including predators. These do, however, during 'lemming years' definitely lay considerably more eggs and rear more offspring. This allows for a more complete utilization of the abundance of the feeding opportunities provided by the rodents during years of a population explosion. But during periods of low numbers of small animals, the broods diminish or reproduction of the birds may stop completely. The high fecundity makes up for the losses occurring due to shortage of food during 'lean' years. Thus, the main characteristic of reproduction in predatory birds on the tundra is not the result of its maxima but of its strong fluctuations in relation to the condition of their food base. This is reflected in adaptation to the ever-changing stock of these birds.

Still another regulating mechanism of population density of predatory birds in relation to the numbers of other birds, consists of the length of the egg-laying period and of the duration of brooding following the laying of the first egg. As a result, the nestlings do not hatch at the same time (the time-span from the appearance of the first to the last of the young ones may be as much as two weeks). In the nests of owls fully feathered large nestlings may be found simultaneously with babies still covered with down. If there is a shortage of food, it is taken mainly by the more active, larger nestling and the younger may perish. This makes it easier to rear the rest of the offspring. Sometimes only two or three out of a large brood of owls or buzzards survive. There are even cases of cannibalism: smaller baby birds may be devoured by the larger during times of starvation.

During years of minimum occurrence of lemmings (depression years) rough-legged buzzards and snowy owls may not nest at all. Over large areas of the tundra it may be possible to observe only single individuals feeding on the remnants of lemmings or birds. The majority of the predators migrate during such years to other areas. For instance, rough-legged buzzards concentrate in the southern parts of the subarctic where in addition to lemmings there are also some species of voles. Skuas switch to eating fish along the sea coast and change their hunting behaviour. The long-tailed skua (*Stercorarius longicaudus*) will consume lots of berries in the southern tundra subzone, whortleberries, cowberries (*Vaccinium uliginosum*, *V. vitis-idaea*) and cloudberries (*Rubus chamaemorus*), as well as insects.

The nesting density of rodentivores is determined by the abundance of lemmings which in turn can be considered as the secondary trophic level in the tundra ecosystem; the latter are the actual consumers of the auto-trophic green plants, which are the primary trophic level. Therefore the rodent-consuming predators usually occupy the tertiary trophic level. Also the majority of other birds can be referred to this level, that is, those consuming invertebrates like the waders, the passerines and the diving ducks, which in turn may become prey to the predators. Only ptarmigan and geese as well as rodents remain typical second level herbivores. But they are only occasionally consumed by predators.

The tundra predators, when switching over to feeding on birds during years of depression in the number of lemmings, become transferred to the fourth level in the food chain. Their productive capacity, that is their ability to increase the stock or to build up the biomass (calculated in units per surface area within an association) is determined by their position in the system of trophic levels. For each additional trophic level a biomass is usually produced which is ten times smaller than that at the level just below. The rodentivorous predators, feeding on lemmings, reach such a density that their numbers can no longer be sustained on the excess of birds. Therefore they cannot maintain their propagation characteristics or the growth rhythm within their population. This causes not only an inevitable restriction in or collapse of propagation but also the death of their offspring and migration to other areas and, as a result, a decline in the density of predators within a given area.

Which is more prevalent on the tundra, carnivory or herbivory?

Ornithologists have repeatedly stated that on the tundra by comparison with the fauna at more southerly latitudes, birds feeding on live food (predators, consumers of fish and insects) predominate. For instance Birulya had already discussed this subject in 1907. This opinion is quite logical since birds, being highly organized animals with more highly developed physiological functions, can less successfully become assimilated into the Arctic because they depend on highly calorific food. In general this condition is the rule. At the same time it is necessary to have reservations when considering tundra birds not only in general but also as groups and in relation to subzones.

Carnivory is characteristic of all groups of birds really thriving at high latitudes. Such are loons (*Gavia*), razorbills (Alcidae), waders (Charadrii) and gulls (Laridae). They are the models of such examples. Herbivorous surface feeding species of true duck (Anatinae) are rather numerous in the forest tundra and in the southern subzone of the tundra (e.g., teal, *Anas crecca*, pintails, *A. acuta*). But there are neither hemiarctic nor hyparctic species among them. They are absent at higher latitudes, being replaced by pochards (Aythyinae) feeding on animal food, such as eiders, scoters (*Melanitta* spp.) or long-tailed ducks (*Clangula hyemalis*), which remain as the representatives of the duck subfamily in the typical and the arctic subzones. The pochards are also represented by a number of typically arctic species among which some are eu-arctic.

There are some interesting data on the relative number of species belonging to different orders. In the Eurasian tundra zone there are about 40% of all waders (snipe, plovers, etc.) in the fauna of the USSR, about 30% of all gulls, 100% of loons, but only 9% of the Gallinaceae and 6% of the small birds. The gallinaceous birds are typical herbivores and among the passerines the majority are also herbivorous although there are some species which are able to consume both plant and animal food.

Making similar comparisons one may run into conflicts also when examining other groups of organisms. For instance all the fishes typical of the high latitudes such as the grayling (*Thymallus arcticus*), the vendace (*Coregonus albula*), the Siberian whitefish (*C. lavaretus pidschian*), the muksun (*C. muksun*), the broad whitefish (*C. nasus*), the whitefish called 'omul', the nelma (*Stenodus leucichtys nelma*), the Siberian sturgeon (*Acipenser baeri*), the perch (*Perca fluviatilis*), the pike (*Esox lucius*), are all typical carnivores. Herbivorous fishes are not found within the tundra zone. Even species using mixed food such as the roach, the dace and the orfe reach only as far as the forest tundra.

Analogous example are furnished by some groups of insects or arachnids. The predatory carabids and staphylinids are the most widely distributed beetles; among the true bugs the coastal predatory species belonging to the family Saldidae reach farthest north. In the subarctic the phytophagous mites diminish sharply although predatory mites and spiders are well represented.

Since the above mentioned examples and figures may not seem convincing, the predominant position of predatory life forms among insects on the tundra makes it necessary to express some essential reservations. Much is still unclear. The predators must of course find food. Although predatory groups seem well represented on the tundra, there must still be a great number of phytophagous species. There is, indeed, a great variation and abundance on the tundra of many typically phytophagous insects, primarily all the leaf miners and sawflies, as well as nematodes consuming both soil algae and dipterous larvae.

However, the bulk of the invertebrates are as a rule not really phytophagous but saprophagous, that is consuming dead parts of plants as well as leaf litter. These include various annelids (earthworms and enchytraeids), mites and dipterous larvae. However, within the enormous group of animals belonging to the saprophages many feed not only on plant debris, but also on bacteria and fungi which, in their turn, also live on vegetable matter. There are data to confirm that groups of soil animals with just such food relationships predominate, but that there are few truly saprophagous species. Thus, feeding on fungi is typical of the springtails. Many nematodes and protozoans also feed on bacteria. Earthworms ingest them together with detritus. Thus, the majority of the tundra invertebrates, making up the basic tier of living matter, are found not at the secondary level but at the tertiary level in the food chain (bacteria and fungi are typical heterotrophs). However, in this case genera are found which also are able to switch over to a higher heterotrophic level. There are, however, many not very clear cases and we risk stepping into a maze of conjecture and opinions, with little foundation. For a thorough and objective analysis of this problem, it is necessary to wait for more exact data concerning the food habits of invertebrates belonging to the saprophagous complex.

Let us return to the birds. From an enumeration of the groups thriving best on the tundra (loons, gulls, Charadrii and snipe as well as diving ducks) it is obvious that all are water birds or waders or belong to groups that act as land birds for only part of the year. All are adapted to feeding on, near or in the water (the skuas, the majority of the Charadrii and gulls). Thus, predominance of carnivory is not characteristic of the entire tundra ornithofauna but primarily of birds living close to bodies of water. All the orders, with the exception of the truly terrestrial birds, are represented by both carnivorous and herbivorous species (gallinaceous birds, owls, diurnal raptors, passerines) on the tundra, at least to some extent (6–10% of the orders within the territory of the USSR). Each of these orders is strongly impoverished in the subarctic but represented by typical hemiarctic or eu-arctic species: the gallinaceous birds by willow grouse and ptarmigan

(*Lagopus l. lagopus, L. mutus*), owls by the snowy owl (*Nyctea scandiaca*), diurnal raptors by the rough-legged buzzard (*Buteo lagopus*) and passerines by the Lapland bunting (*Calcarius lapponicus*), horned lark (*Eremophila alpestris*) and snow bunting (*Plectrophenax nivalis*).

Among the passerines on the tundra, the insectivores dominate in numbers. However, as has been shown above, these have often extended their ranges from south of the subarctic or are related to intrazonal biotopes. The most typical tundra inhabitants, the horned larks, the longspurs and the snow buntings, consume mixed food.

Geese also belong to the order of Anseres, of which we have above mentioned the pochards (Aythyinae) and the surface-feeding ducks (Anatinae). Geese are mainly terrestrial species, feeding on terrestrial plants. No doubt the Anserinae are one of the groups thriving best of all on the tundra and are represented by some ten or eleven species. Thus herbivorous geese and carnivorous diving ducks are most often represented within the tundra zone, with herbivorous surface-feeding ducks occurring to a somewhat lesser extent.

Among the mammals met with in the subarctic, many are predators and, in general, carnivorous, e.g. the wolf (*Canis lupus*), the fox (*Vulpes vulpes*),

Fig. 60. The arctic fox (*Alopex lagopus*), a very typical species among the tundra predators (photo.: A. V. Krechmar).

the wolverine (*Gulo gulo*), the brown bear (*Ursus arctos*), the weasel (*Mustela nivalis*), the stoat (*M. erminea*) and some species of ground squirrel. This has not once been mentioned as one of the characteristic traits of the mammalian fauna of the tundra. However, all the species enumerated are relatively recent arrivals from other zones, invading the southern parts of the tundra. Those typical of the subarctic or distributed over the entire area are rather sporadic. Among the predatory mammals only two are representatives of a truly arctic fauna, the polar fox (*Alopex lagopus*) and the high-arctic polar bear (*Ursus maritimus*). The polar fox is the only indigenous tundra species among the predatory mammals which is of essential importance within the arctic biocoenoses (Fig. 60).

It is, however, among the herbivorous rodents and the ungulates that we meet with the largest numbers of typical tundra endemics. These are eu-arctic species such as the collared and the Siberian lemmings (*Dicrostonyx torquatus, Lemmus sibiricus*) and the musk ox (*Ovibos moschatus*), as well as the hyparctic reindeer (*Rangifer tarandus*), and the narrow-skulled and Middendorff's voles (*Microtus gregalis, M. middendorffii*).

We have here only broached upon the problems belonging to very complicated and badly investigated fields of contemporary ecology, that is the problems of tropho–energy relationships and the correlation between structure and function within the associations. It is not possible here to enter upon these subjects in depth. Let us just mention that the facts presented emphasize the necessity of a strictly concrete analysis of the principles concerning the life of tundra associations. In spite of their poverty as revealed by their simple structure, tundra ecosystems are full of contradictions and peculiarities in every detail and respect. Knowledge of the laws governing the structure of tundra associations and the mutual relationships of their parts is a vital contemporary task. Without an understanding of this, a rational utilization of tundra resources, as well as protection of its delicate environment, will be impossible.

9

Man and the tundra

Ethnographers believe that man colonized the tundra landscape during the postglacial period at the time of the Palaeolithic or the early Neolithic Periods. This occurred about 7–8 thousand years before the present time. All the early history of man in the Eurasian far north was linked to forms of hunting for the reindeer. Fishing, the use of edible plants for food, hunting for game birds and sea animals were clearly of secondary importance during the early stages of the colonization of the tundra. Only the reindeer was able to furnish adequate amounts of food and all the other necessities of life. Using ecological terminology, it could be stated that the ancient peoples of the tundra were 'monophages'.

With time the ability and implements necessary for hunting reindeer were perfected. An especially important stage in the cultural development of the ancient hunters was the invention and use of boats for capturing reindeer while in water during the times of migration across rivers. The evolution of boats is in itself a very important event in the history of Arctic peoples. Great numbers of reindeer were taken during the autumn at the time of migration to their winter pastures in the forest tundra and the taiga and a lesser number during the spring migration to the calving grounds and summer pastures on the tundra. The capture of a large quantity of reindeer during the short period of time during migration necessitated the co-operation of a lot of people, fulfilling many varied tasks. This ability to hunt an abundance of reindeer allowed the people to take a practically unlimited quantity of meat which led to a rapid growth of camps within the tundra zone and attracted different southern tribes which led to the formation of new ethnic groups.

The abundance of reindeer determined for a long time the encampment density of ancient hunters. According to calculations made by Simchenko (1976) the number of wild reindeer making up the tundra populations in

Eurasia was about three million head with an annual increment of about 7%. In relation to this figure the number of ancient hunters may have numbered not more than 11 000 people. The further increase in native colonization of the tundra was connected with a more complete utilization of its other resources such as fish and sea animals and with the development of reindeer herding as well as trade with other more southern peoples.

During colonization of the tundra the archaic life style was preserved for a longer time than among the people in middle latitudes. Ethnographers explain this as a result of the monotony and the stability of their production basis, that is the reindeer.

An ability to take reindeer had become perfected at an early stage of the ancient hunting culture. This was also associated with a prolonged period of standstill in the development of the material culture following the evolution of more rational hunting methods, for example, actually driving the reindeer into water. It was, however, a matter not only of the stability of the production basis as much as of its instability in terms of intensive exploitation.

At the peak of their density in the tundra zone the annual increment of the reindeer herd was not as good but rather below that of ungulates at lower latitudes. Also it is necessary to take into account that the reindeer was not only the basic but the only source of support for the ancient hunters. There was no substitute for reindeer meat. Therefore normal existence was possible only if based on a guaranteed output of an adequate quantity of reindeer, that is by a surplus by comparison with the density of man. Thus, constant conservation by means of low intensity utilization and a comparatively low density of colonization was under such circumstances the only possible mode of maintaining an equilibrium between human society and the surrounding environment. The ancient colonizers were probably not able to exploit the productive base 'to the limit of the possible' or even to exceed by much the limit of a permissible share in the reindeer stock.

In addition to hunting for reindeer, a nomadic lifestyle developed during the colonization of the far north. A type of back-and-forth migration was characteristic, that is, every summer the hunters followed the reindeer to some site where they crossed a river and in the autumn they returned with them to the wintering quarters in the forest tundra and the northern taiga. It is quite evident that the ancient peoples were able to live on the continental tundra only during the summer. It seems unlikely that they were able to store enough meat for the entire winter. Besides, on the tundra it is impossible to find an adequate supply of firewood. Fishing through the ice was still not practised by primitive hunters.

An important milestone was passed during the settling of the subarctic when reindeer were tamed and reindeer husbandry originated. At first reindeer were utilized only in a wild state during the hunting period, but later they were also used for the purpose of breeding. In connection with the development of reindeer husbandry a multitude of new traits appeared in the social structure of tundra inhabitants. In addition the various tribes also perfected methods of hunting other animals and of fishing. This led to a rapid reduction in the number of wild reindeer which had actually started already during Neolithic time.

It is evident that even at an early stage the problem arose as to whether it was more advantageous to hunt for reindeer or to breed them. The question concerning rational correlation between herds of wild and domesticated populations has not been solved even in our time. One can sometimes hear derogatory remarks regarding the 'wild ones' from practising reindeer herders and students. Usually these are blamed for two 'sins': they ruin the pastures and they lure domesticated animals away at times when the herds make contact. Thus, the well known developer of tundra husbandry, V. N. Andreyev, wrote in the journal *Ecology* in 1977 about the wild reindeer and how they wreck reindeer husbandry, stating that they damage the pastures, complicate the maintenance of the domestic herds and cause loss of livestock. In this connection he suggested limiting the number of wild reindeer and moving them to pastures unfit for grazing domesticated herds (i.e., to arctic or alpine tundras).

For those who have been on the tundra where the wild reindeer live and who have studied their pastures and their migration, this problem seems exaggerated. According to Andreyev there are at present about 2.5 million head of domesticated reindeer (over a pasture area with a potential capacity for 3.4 million animals) and 800000 (according to others for *c.* 600000) wild animals. Basically the wild reindeer consist of three herds: one on Kol'skiy Poluostrov under conditions of reservation management, one in Taymyr and one in northern Yamal. The territories occupied by these herds are not very large by comparison with all the land occupied by reindeer husbandry.

The largest one is in Taymyr (see Fig. 49). The sites used for its basic summer pastures and as calving grounds (in the northern half of Taymyr) lie outside the limits of the reindeer husbandry range, that is, where it is not really profitable to graze domesticated animals. Only the wild reindeer are able to successfully utilize the wide ranging but low-productive pastures of these severe high latitude landscapes without essentially causing any real damage to the vegetation. The Plato Putorana where the wild reindeer concentrate during winter are also of little use for utilization by the reindeer industry. Contact between wild and domesticated reindeer in these areas is possible only over relatively short periods of time.

Thus, regulation of the interrelationship between domesticated reindeer and the Taymyr herd may be accomplished in a very simple manner: by an insignificant delimitation of their migration territory. The Taymyr herd, consisting of about 400000 head, is our national pride and joy and an example of successful preservation of a species otherwise subject to intensive eradication. Now the herd is being exploited by hunting management and provides profit for the government. But, without doubt, the wild reindeer ought to be given that small territory in spite of the fact that it might be of use for reindeer husbandry. It contradicts the suggestion once formulated by Andreyev regarding the removal of wild reindeer to sites 'unfit for grazing herds of domesticated reindeer'. By that he meant those arctic and alpine tundras on which the domesticated reindeer would only be able to eke out a pitiful existence.

Hunting for sea animals such as seals, walrus and whales may have developed at a rather late stage. For this purpose large and sturdy boats and powerful harpoons were necessary and the methods of hunting were quite elaborate and required many complicated skills. Sea animals could not provide the ancient people with all their necessities, especially not with light and warm clothing. In addition, these animals are both rare and difficult to catch, especially during winter, in the Kara, Laptev and East Siberian Seas. Only in the warmer Atlantic and Pacific waters are the sea animals able to form an adequate basis for the life of man. It was therefore especially in Chukotka and Greenland that the earliest coastal settlements were formed.

Consumption of vegetable food in the far north was associated with the hunting of sea animals. The meat of the walrus (*Odobenus rosmarus*) and of seals (Phocidae) is very greasy and its chemical composition is comparatively poor. Thus, for normal digestion and assimilation of fatty food it is necessary to add carbohydrate and a variety of inorganic salts. When feeding on the fat meat of sea animals, people felt it necessary to use plant food as a supplement to providing the various substances required and as a kind of fermented seasoning. In this connection the use of plants in their diet was broadly practised by coastal people, e.g., the Chukchi and the Eskimaux. The use of herbs in the food and the tradition of collecting plants became perfected over the centuries. In the graves of Eskimau women from the second millenium B.C. bone hoes have been found which are similar to those used for cultivating plants in our own time.

The Chukchi use about 20 different species of wild plants in their diet.

They boil them as a condiment in soups, use them fresh, often mixed with blubber, and store them in dry or fermented form (Sokolova, 1961). The plants most frequently used by them are a saxifrage (*Saxifraga nelsoniana*), mountain sorrel (*Oxyria digyna*), arctic dock (*Rumex arcticus*), snakeweed (*Polygonum ellipticum*), roseroot (*Rhodiola atropurpurea*), the leaves of some willows (*Salix pulchra, S. arctica*) and flowers of the milk-vetch (*Astragalus umbellatus*). They also collect a marine brown alga (*Laminaria*). The Chukchi women are skilled in distinguishing all these plants and know the best time for collecting the different species.

However, many people in the tundra zone do not eat much plant food and, evidently, do not feel like using this resource. This is most common among those feeding mainly on reindeer meat. This meat is quite lean but very rich in different chemical substances. The first time those who try a soup made from fresh summer reindeer meat are surprised by its delicious smell. It is reminiscent of the scent of mushrooms or a fragrant herbal brew. The contents of the stomach are also used.

Spontaneous interrelationship between man and the plant kingdom in the far north is rather limited. There are no forms of activity there which lead to direct and often irreversible changes in vegetation cover such as, e.g., ploughing, mowing or cutting down forests. However, the effect of man on the plant world of the tundra takes many other forms. In 1883 the well known investigator of arctic vegetation, F. Kjellman, wrote:

Against his will and, indeed, quite without knowing it, the Chukchi have become cultivators of plants. Almost everywhere around his tent species of plants grow in great profusion, some of which may be entirely absent in the neighbourhood, others encountered in very small numbers or growing widely dispersed. There is no doubt that some of them arrived here without direct participation of the Chukchi. But they found a favourable environment for growing on the dunghills, which in time arose around the campsites, and then became established and propagated there. And no doubt others owe their presence to the Chukchi who had collected them far away but then threw some away in the camp as a kind of refuse, whereafter they took root and propagated. (Quoted from Dorogostayskaya, 1972.)

Changes in the vegetation cover in places where man has settled within the tundra and his introduction of more southerly species constitute very interesting ecological problems. Especially important is the question regarding the formation of weed complexes at the northern outposts of agriculture. This problem has for several years been studied by Dorogostayskaya, who has published a series of special monographs on it. She worked on the classification of the anthropophilic (i.e., associated with man) flora of the tundra. She distinguished only two large groups: the apophytes, i.e., local species becoming established in settled areas and disrupting existing biotopes by propagating more successfully than in their own, original native habitats, and the anthropochores, i.e., species brought into the tundra zone by man.

What kind of species have become apophytic within the tundra zone? An opinion exists that vegetation in anthropogenic habitats (which could also be called ruderal sites or dumps, often overgrown with nettles or burdock (*Urtica, Arctium*), etc.) formed originally in zoogenic habitats. By this is understood a basically similar type of effect on the environment caused by both animals and men: disruption of the vegetation cover, tearing up the ground and enriching it with nutrient matter, especially nitrogen compounds, etc. Plant associations in the neighbourhood of settlements are, thus, always similar to those, for example, seen around fox dens, on bird cliffs and in places of intensive rodent activity.

The apophytic flora may be distinguished into nitrophilic (preferring a surplus of nitrogen) and erosiophilic (associated with broken up soils and a disrupted vegetation cover) types.

Typical nitrophilic species in the tundra are the arctic buttercup (*Ranunculus hyperboreus*), the alpine foxtail grass (*Alopecurus alpinus*), the arctic ragwort (*Senecio arcticus*), and another grass (*Arctophila fulva*). These are especially common on polluted sites, often forming a dense growth on muddy sites and on dumps. The erosiophiles are under natural conditions associated with loose, sandy or gravelly, erosion-prone ground. In such habitats they gain a foothold thanks to their ability to grow fast by means of rhizomes or sturdy roots, resistant to the mechanical effect of mobile soils. A wormwood (*Artemisia tilesii*) is typical of apophytic–erosiophilic plants. It is common in various sites with crumbling soils.

The majority of erosiophils are at the same time also nitrophilic. Such species are most numerous in anthropogenic habitats. The most typical representatives are the large-flowered mayweed (previously known under the name of *Matricaria grandiflora* but now named *M. maritima* spp. *phaeocephala*). It forms a dense growth in and around arctic settlements, is common on sites with ruins of buildings, on overgrown burrows, in muddy spots, etc. It is distinguished by its abundant flowering. As nitrophilic–erosiophilic representatives may also be counted such species as marsh ragwort (*Senecio palustris*), arctic and alpine poas (*Poa arctica, P. alpina*) arctic scurvy-grass (*Cochlearia arctica*, a crucifer), the common horsetail (*Equisetum arvense*), the tansy-mustard (*Descurainia sophia*) and the tansy (*Tanacetum bipinnatum*).

Tundra apophytes are mainly species of natural intrazonal habitats (eroded slopes, beaches, nival meadow-like sites, etc.). Among them there are almost no species characteristic of zonal plakor associations on level, moderately wet, well-drained, loamy soils. The disposition to colonize anthropogenic biotopes is typical of some moisture-loving plants such as the Lapland butterbur (*Nardosmia frigida*), a typical inhabitant of peaty and boggy sites as well as river banks (similar examples are also found for insects).

On trampled and manured sites near exits from barns or houses, where the natural tundra sward has been torn up or burnt away, a special variety of vegetation has developed reminiscent of common ruderal growth at middle latitudes. Other sites are found, which could be called 'improved tundra', that is carrying a vegetation cover analogous to that of a meadow. Such a vegetation can be encountered near some settlements and on sites of former camps, temporarily used by both hunters and fishermen. Grasses forming a meadow-like sward are common but there are few mosses because these have been burnt away or become completely eradicated due to pollution.

The well-known enthusiast promoting colonization of the north, A. V. Zhuravskiy, considered this a suitable method for increasing the biological productivity of tundra and he made much propaganda of it. In a paper with the enticing title 'Arctic Areas in a New Light' (1915), he wrote:

The tundra is, consequently, a type of Russian wasteland which does not involve a specific 'poverty of the soil' since the flora and fauna of the mosses and lichens are basically identical to those in the native forests dependent on a mossy covering; what is more, these mosses which do not utilize the soil products of oxidation and decomposition, thus, do not and cannot define the composition of the soil merely by their presence on it; it may even be said that they 'preserve it' . . . Be that as may be, everything, without exception, is a fact, based on actual materials concerning all groups of fauna, flora, geology, ethnography, and my own observations force me to state that agriculture, animal husbandry and the production of oil are geographically specific branches quite possible to manage in the northlands and mainly within the 'arctic' north.

Zhuravskiy was clearly carried away and grossly overestimated the possibilities of methods for achieving a real increase in the productivity of the tundra. Since then, practical results of experiments which give hope for an improvement of the native associations have been obtained only within the forest tundra, e.g., in the Salekhard area. On sites with bush–moss associations, meadows of hay have been established yielding an annual crop of 30–40 centners per hectare [*c.* 1500–2000 kg/hectare] (Filippova 1963). The outlook for creating a grazing economy in this area seems to have been proven, but within the typical and even the southern tundra subzones this is still a hope for the future. Improvements can be achieved there only by draining raised surfaces where natural vegetation lacks in common herbs or grasses.

Grass-apophytes are very sensitive to the extent of disruption of the biotopes. They will take hold only under conditions where the mosses have been destroyed but a peaty horizon has been preserved. On habitats which have been ploughed and well cultivated for a long time, tundra apophytes, as a rule, do not thrive. Dorogostayskaya demonstrated that in northern outposts of agriculture hardly a single species from the local flora had invaded ploughed areas, on which only plants or weeds brought in from the south had established themselves. This fact bears indirect witness to the opinion that tundra apophytes will be of only limited importance in the process of improving and cultivating the tundra. It is necessary to keep in mind that mosses are strongly competitive with grasses in arctic areas.

There are also different species of anthropochores associated with the colonization of man, among which some groups may be distinguished according to their degree of incorporation into anthropogenic biotopes: a temporary group, that is, composed of species met with at some distance from their natural habitat and not adapted to typical tundra conditions, successfully colonizing only strictly limited and specific biotopes (e.g., greenhouses, hotbeds), and a naturalized group, that is species which have already become established in the northlands and commonly grow in various anthropogenic biotopes. The pathways by which the anthropochores are introduced into the tundra differ also: by intentional importation of cultivated species and by unintentional introduction, for example, mixed in with seed material and food products.

Dorogostayskaya describes cases which demonstrate that southern plants have been brought to arctic areas in some quantity. Around Noril'sk, on sites where hayfields have been ploughed, she observed 19 species of plants which were growing some hundreds to a thousand kilometres from their native habitats, although only in a vegetative form. However, only very few anthropochores are able to acclimatize to any important extent within the Arctic. More often they last for a short time only in the various biotopes but do not produce mature seeds and soon die off. Only in the southern parts of the subarctic have some of them successfully colonized anthropogenic habitats. Thus in some areas within the forest tundra and in the southern tundra it is possible to find in settled areas such plants as the common nettle (*Urtica dioica*), plantain (*Plantago major*) and knotgrass (*Polygonum aviculare*), etc. However, the acidic conditions characteristic of the tundra and especially of the typical tundra subzone constitute an insurmountable barrier for these species.

Individual species display great activity when invading anthropogenic biotopes in the subarctic. As an example we may mention that the pineapple weed (*Chamomilla suaveolens* = *Matricaria matricarioides*), which was brought into Vorkuta some years ago, has produced a tremendous number of offspring forming a dense growth there. Another example is the common groundsel (*Senecio vulgaris*), which has also spread quickly within tundra settlements.

Some of the anthropochorous plants have invaded various intrazonal biotopes, e.g., along shorelines, on sand bars, precipices and among boulders. In such sites there is no competition from native tundra species and they have often succeeded in establishing themselves. Examples are the tansy (*Tanacetum vulgare*), the bromegrass (*Bromopsis inermis*), wormwood (*Artemisia vulgaris*) and the pineapple weed (*Chamomilla suaveolens* = *Matricaria matricarioides*), etc. However, in zonal tundra biotopes, these anthropochorous species are unable to get a foothold.

Similar regularities occur also when analysing the formation of animal colonization of anthropogenic biotopes. Here the complex of so-called synanthropic flies (i.e., flies associated with man) is a good example. The subarctic fauna of these types of animals can be divided into a few groups reflecting basic modes of formation: typical synanthropes, entering the tundra zone only in association with man (ordinary house-flies, small-size house-flies, another fly named *Hydrothaea dentipes*, common yellow dung fly, etc.); those widely distributed also in the subarctic to which belong the carrion flies (*Protophormia terraenovae* and *Cynomyia mortuorum*); and such arctic species which to a certain extent have a tendency to inhabit human settlements, mainly in association with refuse.

The specific natural conditions of the subarctic are such that none of the typical synanthropes (i.e., those belonging to the first group) are able to propagate to any important extent within the arctic settlements. The most numerous synanthropous flies in the tundra zone belong mainly to the second group but there are also individual species which are typically eu-arctic, for example the blue carrion fly (*Boreellus atriceps*) and a small fly propagating on various kinds of refuse, *Neoleria prominens*.

A similar relationship can be found amongst birds. The white wagtail (*Motacilla alba*) and the wheatear (*Oenanthe oenanthe*) are common in settled areas all over the tundra. They are otherwise widely distributed inhabitants of intrazonal biotopes. Typical synanthropous birds, brought unintentionally into the high latitudes by man such as tree sparrows (*Passer montanus*), swallows (*Hirundo rustica*) and starlings (*Sturnus vulgaris*), are limited to settlements in the southern parts of the subarctic. An important number of species belonging to the indigenous fauna has to a certain extent shown a tendency toward synanthropism but of these only snow buntings

(*Plectrophenax nivalis*) have become constant inhabitants, occurring in great numbers within the arctic settlements. They replace the sparrows.

The examples mentioned above bear witness to the fact that the effect of man on widely differing groups of organisms is subject to similar interrelationships as in nature. It is possible to notice several special premises regulating man's relationship with the animal and the plant world of the tundra.

The biological resources which allow a low variety of life-styles to develop on the ordinary lowlands of the far north are fairly abundant. The native resources to be utilized by man may be more abundant here than in landscapes situated further south. However, this is not the result of high productivity, that is annual increment, but the consequence of an accumulation of populations or material depending on low intensity of undisturbed biological processes. It depends also on the concentration within the basic portion in the productivity chain of a small number of species at different trophic levels. Just because of the apparent richness of the resources in the north, they risk being quickly exhausted and only slowly restored. There is, it seems, a paradoxical situation in the Arctic: at a very low population density of humans, many species of animals were more rapidly eradicated than when there are relatively more people. It is enough to mention that animals such as the great auk (*Alca impennis*) is already extinct and the walrus (*Odobenus rosmarus*), the polar bear (*Ursus maritimus*) and the musk ox (*Ovibos moschatus*) are threatened with eradication. Even the ancient inhabitants of the Eurasian tundras, the mammoth (*Mammuthus primigenius*), were most probably eradicated by a low number of ancient people.

Because of this, the mastering of the north by man had at the start a very passive character, being more extensive than intensive. People wanted to move in, to master new areas, but were not able to make any substantial inroads into the resources exploited and did not succeed in making any large or lasting yields materialize from the resources utilized. The economic structure of many northern peoples was such that it has to be acknowledged that under the local conditions people were unable to exploit fully the stock of animals but were themselves subjected to their mode of life. For the Nganasani, a Siberian people, the relationship between the members of the tribe and their families, the choice of a camp site and the distribution of the work load were entirely regulated by the life cycle and the seasonal migrations of their herds of reindeer.

The characteristics typical of tundra animals mentioned previously necessitate specific measures for their utilization as well as for their protection. We must have a carefully guarded approach to and a more reasonable exploitation of the natural resources. One of the routes toward a sensible exploitation of tundra animals is extensive: an even distribution of productive output over the entire territory, strict alternation when exploiting the offspring, with occasional periods when the populations are 'rested'. We must have a careful management of the utilization of arable land, etc.

The very sharply differentiated types of landscapes are the most important characteristics of arctic associations with a direct relationship to practical utilization and protection of resources. It has been demonstrated above that the important gradients in the climate here appear over much shorter distances than at more southerly latitudes. Because of this the composition of the flora and the fauna and the types of associations extending over the tundra zone from its northern to its southern limits change there very rapidly. The three tundra subzones (the arctic, the typical and the southern) differ from each other in important characteristics. For instance, animals dominant in arctic tundras are hardly met with in the southern subzone and vice versa. All this necessitates a strict differentiation in approach both to the associations and to the landscape as a whole. In addition to zonation within the Arctic there is also a very strong sectorial differentiation of the landscapes. The Atlantic and Pacific coastlands and the Siberian tundras are all very different from each other in respect to structure of associations and composition of flora and fauna.

Before utilizing the arctic resources yet another important aspect must be considered. Its natural life has indeed a peculiar 'resonance' (Kryuchkov, 1976). Often apparently minor infringements here become magnified and lead to irreversible basic changes in associations and landscapes. Thus, minor damage can lead to a reduction of the permafrost, cause landslips and promote the appearance of thermokarst formation. A slight lowering of the temperature may even lead to a local glaciation.

Analogous phenomena also have an effect on tundra associations. The importance of superdominants within their structure is special; that is where a single species (or a group of plants or animals) plays a very important role for the existence of many other components within the associations. The lemmings are a good example: they are the basic food source for all the tundra predators; another is the crane-fly, or rather two or three species of these tipulids, which are the basic food of a great many insectivores on land, and the chironomid larvae, the basic food for aquatic vertebrates. A disruption of the populations of such species or groups will inevitably simultaneously provoke both irreversible and negative changes in the populations of many other species.

The relatively simple structure of the arctic associations and their strong dependence on a great number of components belonging to single basic units cause a rapid diffusion within the associations of, for instance, poisonous materials or radionuclides. Research concerning the consequences of an application of insecticides in the fight against blood-sucking gnats has shown that the poisons very quickly disperse among the basic members of the tundra associations, follow along the food chain and turn up within a short time in the bodies of different vertebrates, lemmings, reindeer, birds and plants, etc. (Shilova, Denisova *et al.*, 1973). At the same time the removal of poisonous material from the ecosystem and processes of autopurification under conditions in the tundra are very limited. It is evident that the application of poisonous material under the extreme conditions present in the northlands requires great prudence.

The vulnerability of nature in the north, its capacity for reacting sensitively to all effects upon it and for magnifying them makes it necessary to establish immediately a system of regulations which are both very specific and well differentiated as well as directed toward maintaining the delicate equilibrium within the ecosystems of the Arctic and the subarctic.

The development of the national economy within the USSR depends to a great extent on resources in Siberia and in the northlands. Hence knowledge of these territories with respect to the economy of this country must be developed steadily. It is necessary to institute wide-ranging, thorough and multifaceted research concerning the effects of man's industrial activities on the natural processes taking place in arctic areas.

It is especially important for conditions in the north to have detailed regulations regarding the principles of coping with the resources while taking into account all possible consequences since nature there is extremely sensitive to the impact of man.

Prominent people as well as councillors, investigators of the north and managers of scientific establishments have drawn up a number of recommendations, directed toward the protection of the environment, and presented plans for future research on these problems. They drew the attention of corresponding departments to requests for organizing a number of reservations and restricted areas in the north, for providing purification of sewage in cities, for bringing under control the application of poisonous materials, for creating new means of transport, for diminishing the destruction of the soil cover, and for enacting complex medical–biological programmes dealing with the problems of man's adaptation to the north.

Immediate recommendations concerning problems of rational utilization of natural resources in the northlands are also found in the long-term programmes called 'Siberia', worked out by the Siberian Departments of the USSR Academy of Sciences. This programme includes references to the main problems of defining the complex management of mining, land, forest and water resources. Special attention is also given to problems dealing with protection of the environment. An important part of this programme is devoted to working out scientifically-based enactments concerning the protection of the biosphere in areas where there is intensive development such as mining, industries and major construction. It provides for new means of protecting the biotopes from pollution and technological violation as well as for methods of rational utilization, restoration and protection of the plant and animal worlds in Siberia and in the north as a whole.

The important general problem of utilizing biological resources is also of direct concern for future management of the north. Since ancient time the activities of man in the arctic areas have been based on hunting wild animals. However, it has now become more and more evident that the time has come when indiscriminate removal of 'necessary' products from the biosphere is no longer possible. In order to be able to remove something from nature, it is also necessary to take care to restore the resource in question. This may be expressed by the words: transition from hunting to management has become necessary. The time has come when man must stop preying on animals and begin breeding and managing them. By this is understood the semi-restricted and fully restricted utilization of productive animals and their complete management: the introduction of aquaculture, the creation of substantial fish-breeding complexes and hatcheries for raising the young of various fish species, and stations for studying the selection and genetics of animals, etc. For a solution to these problems complicated and comprehensive research procedures are required in terms of propagation, growth and development and the study of the trophic relationships of various organisms under typical arctic conditions. It is also necessary to work out acclimatization and re-acclimatization procedures over wide ranges.

A very important problem concerns cultivation of land in the far north, for instance, restoration of strongly depleted and overgrazed reindeer pastures and of the soil cover in places where there are construction and mining enterprises. Detailed research is required concerning the course of ecological succession within the subarctic ecosystems. Only a thorough knowledge of the procedure of successional flow under extreme conditions will allow us realistically to direct the restoration processes.

Management of the northlands involves also a great number of medical and biological problems. The life of man under extreme conditions in polar lands is accompanied by intense adaptive processes in his body. In this connection it seems to be basically more important to establish methods for selection of professional people while working out optimum prophylactic measures. In the future it will be necessary to develop research into many aspects of the ecology of man under conditions in the Arctic and subarctic. Within the territory of the Arctic to which the tundra belongs attempts are being made to locate the centres of diseases such as tularaemia and leptospirosis. It is necessary to study how opportunities arise for the spread of these diseases, which could affect the future growth of human population density. It is also necessary to work out prophylactic methods to treat them. Obviously it is even more necessary to study every kind of hygienic and ecological criterion for projected industrial complexes within the arctic territory.

Finally, when utilizing resources in the Arctic which affect its nature, it is necessary to keep in mind that the Arctic has a tremendous impact on climate and weather all over the world. The level of the world's oceans depend on the status of sea ice and glaciers. A change in the temperature regime within the arctic territories will be sharply reflected in atmospheric circulation and humidity conditions in more southerly areas and can give rise to drought or periods of excessive precipitation at temperate latitudes. Man's tampering with arctic landscapes, its waters, soils, permafrost conditions and its atmosphere can therefore have all-encompassing and unfortunate consequences for our entire planet. Therefore it is important for the scientists of the future to perfect the very complex system of managing the natural resources of this land.

Bibliography

Aleksandrova, V. D. (1961). Vliyaniye snezhnogo pokrova na rastitel'nost' v arkticheskoy tundre Yakutii. [Importance of the snow cover over the vegetation in arctic Yakutia.] In the symposium: *Materialy po rastitel'nosti Yakutii*. [Material on the vegetation of Yakutia.] Leningrad.

Aleksandrova, V. D. (1977). *Geobotanicheskoye rayonirovaniye Arktiki i Antarktiki*. [The Arctic and Antarctic: their division into geobotanical areas. English translation, Cambridge, 1980.] Leningrad.

Andreyev, A. V. (1975). Zimnyaya zhizn' i pitaniye tundryanoy kuropatki na kraynem severo-vostoke SSSR. [Winter life and diet of the willow grouse in the far north-eastern SSSR.] *Zoologicheskiy Zhurnal*, **54** (5).

Bol'shakov, V. N. (1972). *Puti prisposoblenia melkikh mlekopitayushchikh k gornym usiloviyam*. [Pathways of the adaptation of small mammals to alpine conditions.] Moscow.

Birulya, A. A. (1907). Ocherki iz zhizni ptits polyarnovo poberezh'ya Sibiri. [Essays on the life of birds along the arctic coast of Siberia.] *Zapiski Academii Nauk*, ser. 8, **18** (2).

Budyko, M. I. (1948). O zakonomernostyakh poverkhnostnogo fiziko–geograficheskogo protsessa. [On the regularity of physico–geographical processes.] *Meteorologiya i Gidrologiya*, 1948 (4).

Chernov, Yu. I. (1966). Kompleks antofil'nykh nasekomykh v tundrovoy zone. [The complex of anthophilic insects in the tundra zone.] *Voprosy Geografii*, no. 69.

Chernov, Yu. I. (1967). Troficheskiye svyazi ptits s nasekomymi v tundrovoy zone. [The trophic relationship between birds and insects within the tundra zone.] *Ornithologiya*, no. 8.

Chernov, Yu. I. (1975). *Prirodnaya zonal'nost' i zhivotnyy mir sushi*. [Natural zonation and animal life on the continent.] Moscow.

Chernov, Yu. I. (1978). Antofil'nyye nasekomye v podzone tipichnykh tundr zapadnogo Taymyra i ikh rol' v opylenii rasteniy. [Anthophilic insects in the typical tundra subzone of western Taymyr and their role in the pollination of plants.] In the symposium: *Struktura i funktsii biogeotsenosov taymyrskoy tundry*. [Structure and function of the biogeocoenoses in the Taymyr tundra.] Leningrad.

Danilov, N. N. (1966). *Puti prisposobleniya nazemnykh pozvonochnykh zhivotnykh k usloviyam sushchestvovaniya v Subarktike. Ptitsy*. [Pathways of adaptation of vertebrates living underground to the conditions existing in the Subarctic. The birds.] Sverdlovsk.

197

Dorogostayskaya, Ye. V. (1972). *Sornye rasteniya Kraynego Severa SSSR.* [Weed plants in the far north of the USSR.] Leningrad.

Dunayeva, T. N. (1948). Sravnitel'nyy obzor ekologii tundrovykh polevok polyostrova Yamal. [Comparative survey of the ecology of tundra voles on the Yamal peninsula.] *Trudy instituta geografii AN SSSR,* no. 41.

Dunayeva, T. N. (1978). K ekologii lemmingov v severo-zapadnoy Yakutii. [On the ecology of the lemmings in north-eastern Yakutia.] *Byulleten' Moskovskogo Obshchestva Ispytatelly Prirody. Otdeleniye Biologicheskoye,* **83** (2).

Filippova, L. N. (1963). Problemy zaluzheniya tundry v Yamalo–Nenetskom natsional'nom okruge. [Problems of the lure of the tundra in the Yamal–Nenets national district.] *Trudy Nauchno-issledovatel'skogo Instituta Sel'skogo Khozyaystva Kraynego Severa,* vol. 11. Norilsk.

Fridolin, V. Yu. (1936). Zhivotno-rastitel'noye soobshchestvo gornoy strany Khibin. [Animal and plant associations in the mountain areas of Khibiny.] *Trudy Kol'skoy bazy AN SSSR,* no. 69.

Gorodkov, B. N. (1930). Vechnaya merzlota i rastitel'nost'. [Permafrost and vegetation.] In the symposium: *Vechnaya merzlota.* [Permafrost.] Leningrad.

Gorodkov, B. N. (1938). Rastitel'nost' Arktiki i gornykh tundr SSSR. [The vegetation of the arctic and mountain tundras of the USSR.] In: *Rastitel'nost' SSSR,* vol 1 (The vegetation of the USSR, 1.] Moscow, Leningrad.

Gorodkov, B. N. (1950). Moroznaya treshchinovatost' gruntov na Severe. (Frost-cracking of the ground in the Far North.] *Izvestiya Vsesoyznogo Geograficheskogo Obshchestva,* **82** (5).

Govorukhin, V. S. (1960). Pyatnistyye tundry i plikativnyye pochvy Severa. [Spotted tundra and heaving soils in the far north.] *Zemlevedeniye* 1960 (5).

Grigor'yev, A. A. (1956). *Subarktika.* [The subarctic.] Moscow.

Kalesnik, S. V. (1970). *Obshchiye geograficheskiye zakonomernosti Zemli.* [General geographical regularities of the Earth.] Moscow.

Karavayev, M. N. (1958). *Konspect flory Yakutii.* (Conspectus of the flora of Yakutia.] Moscow, Leningrad.

Kerner, A. (1930). *Zhizn'rasteniy.* [Plant life.] St Petersburg.

Kishchinskiy, A. A. (1971). Sovremennoye sostoyaniye populyatsii dikogo severnogo olenya na Novosibirskykh ostrovakh. [Present state of the populations of wild reindeer on the Novosiberian Islands.] *Zoologicheskiy Zhurnal,* **50** (1).

Krechmar, A. V. (1966). Ptitsy zapadnogo Taymyra. [The birds of western Taymyr.] *Trudy Zoologicheskogo Instituta AN SSSR,* vol. 39.

Kryuchkov, V. V. (1976). *Chutkaya subarktika.* (The delicate subarctic). Moscow.

Lozina-Lozinsky, L. K. (1972). *Ocherki po kriobiologii.* [Essays on cryobiology.] Leningrad.

Matveyeva, N. V. and Chernov, Yu. I. (1977). Arkticheskiye tundry na severo-vostoke Taymyra. [The arctic tundras in north-eastern Taymyr.]. *Botanicheskiy zhurnal,* **62** (7).

Middendorff, A. F. (1860–7). *Puteshestviye na sever i vostok Sibiri.* [Travels in northern and eastern Siberia.] St Petersburg.

Mil'kov, F. N. and Gvozdetskiy, N. A. (1976). *Fizicheskaya geografiya SSSR.* [Physical geography of the USSR.] Moscow.

Osmolovskaya, V. I. (1948). Ekologiya khishchnykh ptits poluostrova Yamal. [Ecology of predatory birds on the Yamal peninsula.] *Trudy instituta geografii AN SSSR,* no. 41.

Parinkina, O. M. (1973). Biologicheskaya produktivnost' bakterial'nykh soobshchestv tundrovykh pochv. [Biological productivity of bacterial associations in the tundra soil.]

In the symposium: *Biogeotsenozy taymyrskoy tundry i ikh produktivnost'*.
[Biogeocoenoses of the Taymyr tundra and its productivity.] Vol. 2. Leningrad.

Pavlov, B. M. (1974). Belaya i tundryanaya kuropatki Taymyra. [Ptarmigan and willow grouse in Taymyr.] Doctoral thesis. Moscow.

Polozova, T. G. and Tikhomirov, B. A. (1971). Sosudistyye rasteniya rayona taymyrskogo statsionara. [Vascular plants in the area of the Taymyr research station.] In the symposium: *Biogeotsenozy taymyrskoy tundry i ikh produktivnost'*. [Biogeocoenoses of the Taymyr tundra and its productivity.] Leningrad.

Rikhter, G. D. (1945). *Snezhnyy pokrov, yego formirovaniye i svoystva.* [The snow cover, its formation and peculiarities.] Moscow, Leningrad.

Ryabchikov, A. M. (1972). *Struktura i dinamika geosfery i yeye yestestvennoye razvitiye i ismeneniye chelovekom.* [Structure and dynamics of the geosphere, its natural development and changes due to man's actions.] Moscow.

Sdobnikov, V. M. (1953). *Po arkticheskoy tundre.* [On the arctic tundra.] Moscow.

Semyenov-Tyan-Shanskiy, O. I. (1977). *Severnyy olen'.* [The reindeer.] Moscow.

Shamurin, V. F. (1966). Sezonnyy ritm i ekologiya tsveteniya rasteniy tundrovykh soobshchestv na severe Yakutii. [Seasonal rhythm and ecology of flowering plants in the tundra associations of northern Yakutia.] In: *Prisposobleniye rasteniy Arktiki k usloviyam sredy.* [Adaptation of plants in the Arctic to the environmental conditions.] Moscow, Leningrad.

Shilova, S. A., Denisova, A. V. *et al.* (1973). O posledeystvii primeneniya insektitsidov v Subarktike. [On the after-effects of using insecticides in the subarctic.] *Zoologicheskiy Zhurnal,* **52** (7).

Shvarts, S. S. (1963). *Puti prisposobleniya nazemnykh pozvonochnykh zhivotnykh k usloviyamy sushchestvovaniya v Subarktike.* Vol. 1. *Mlekopitayushchiye.* [Pathways of adaptation of vertebrate animals living in the ground to the conditions in the subarctic. Vol. 1. The Mammals.] Sverdlovsk.

Shvetsova, V. M. and Voznesenskiy, V. L. (1970). Sutochnyye i sezonnyye izmeneniya intensivnosti fotosinteza i nekotorykh rasteniy Zapadnogo Taymyra. (Daily and seasonal variation in intensity of the photosynthesis of some plants in western Taymyr.] *Botanicheskiy Zhurnal,* **55** (1).

Simchenko, Yu. B. (1976). *Kultura okhotnikov na oleney severnoy Evrazii.* [The culture of reindeer hunters in northern Eurasia.] Moscow.

Sokolova, T. G. (1961). K voprosu ob ispol'zovanii chukotskim naseleniyem dikoy flory v rayone mysa Dezhneva. [On the problem of wild plants utilized by the inhabitants in the area of Cape Deshnev.] *Zapiski Chukotskogo Krayevedcheskogo Muzeya.* Magadan.

Strel'nikov, I. D. (1940). Znacheniye solnechnoy radiatsii v ekologii vysokogornykh nasekomykh. [The importance of solar radiation for high alpine insects.] *Zoologicheskiy Zhurnal,* **19** (2).

Syroyechkovskiy, Ye. V. (1978). Razmeri lebedey, gusey i kazarok v svyazi s adaptatsey k polyarnym usloviyam. [Dimensions of swans, geese and brants and the relationship to adaptation to arctic conditions.] *Zoologicheskiy Zhurnal,* **57** (5).

Takhtadzhyan, A. L. (1978). *Floristicheskiye oblasti Zemli.* [The floristic regions of the world.] Leningrad.

Tikhomirov, B. A. (1959). *Vzaimosvyazi zhivotnogo mira i rastitel'nogo pokrova tundry.* [The interrelationship between animal life and plant cover on the tundra.] Moscow, Leningrad.

Tikhomirov, B. A. (1963). *Ocherki po biologii rasteniy Arktiki.* (Essays on the biology of the plants in the Arctic.] Moscow, Leningrad.

Tolmachev, A. I. (1974). *Vvedeniye v geografiyu rasteniy.* [Introduction to phytogeography.] Leningrad.

Tugarinov, A. Ya. (1928). O proiskhozhdenii arkticheskoy fauny. [On the origin of the arctic fauna.] *Priroda*, 1928 (7–8).

Vangengeym, E. A. (1977). *Paleontologicheskoye obosnovaniye stratigrafii antropogena Severnoy Azii.* [Paleontological basis for an anthropogenic stratigraphy in northern Asia.] Moscow.

Washburn, A. L. (1958). Klassifikatsiya strukturnykh gruntov i obzor teoriy ikh proizkhozhdeniya. [Classification of patterned ground and a survey of its origin.] In: *Merzlyye gornyye porody Alyaski i Kanady.* [Types of frozen ground in Alaska and Canada.] Moscow.

Yudin, B. S., Krivosheyev, V. G. and Belyayev, V. G. (1976). *Melkiye mlekopitayushchiye severa Dal'nego Vostoka.* [Small mammals in the northern far east.] Novosibirsk.

Yurtsev, B. A. (1966). Gipoarkticheskiy botaniko-geograficheskiy poyas i proizkhozhdeniye yego flory. [The hyparctic botanical–geographical belt and the origin of its flora.] *Komarovskiye Chteniya XIX.* [*Komarov lecture* no. 19.]. Moscow, Leningrad.

Yurtsev, B. A. (1974). *Problemy botanicheskoy geografii severo-vostochnoy Azii.* [Problems of geobotany in north-eastern Asia.] Leningrad.

Zhitkov, B. M. (1913). Poluostrov Yamal. [The Yamal peninsula.] *Zapiski Russkogo Geograficheskogo Obschestva*, 1913.

Index